Risk Management for Water and Wastewater Utilities

Risk Management for Water and Wastewater Utilities

Second Edition

Edited by
Simon J. T. Pollard and Tom Stephenson

Published by **IWA Publishing**
Alliance House
12 Caxton Street
London SW1H 0QS, UK
Telephone: +44 (0)20 7654 5500
Fax: +44 (0)20 7654 5555
Email: publications@iwap.co.uk
Web: www.iwapublishing.com

First published 2016

British Library Cataloguing in Publication Data
A CIP catalogue record for this book is available from the British Library

ISBN: 9781780407470 (Hardback)
ISBN: 9781780407487 (eBook)

Contents

Introduction to the Water and Wastewater Process Technologies Series

Unit operations are linked together in a flowsheet to provide water and wastewater treatment.

This series of texts, termed *modules*, is designed to provide an education in water and wastewater treatment from a process engineering perspective. At the end, you should have a thorough understanding of the design, operation and management of water and wastewater treatment processes. This might be the treatment of raw water to a potable standard, of an industrial effluent to a standard acceptable for disposal to sewer, or treatment of municipal sewage to meet environmental discharge consent. In almost all cases, a single 'process', usually termed a *unit operation*, will not be able to achieve the required level of treatment. Treatment is achieved through linking the right unit operations in a *flowsheet.* This represents a multi-barrier approach to the treatment of raw water to produce potable water, or for removal of impurities in wastewater for acceptable discharge to receiving water bodies.

With the emphasis on utility infrastructure and environmental impacts, rather than the processes, technologies required for the provision of potable water and for the treatment of wastewaters have been taught separately. Teaching tends to follow an 'application down' approach. Elements of training at almost all levels of the water sector are usually presented in this way. For example, a programme on wastewater treatment might be structured with courses entitled *Introduction to municipal wastewater treatment; Advanced municipal wastewater treatment; Low cost sewage treatment; Industrial effluent treatment;* and so on. For drinking water, it is likely to be an *Introduction to potable water treatment*, followed by *Advanced potable water treatment*, etc. This means that unit operations relying upon different scientific principles are taught together, arguably making it difficult to instil a true understanding of the principles governing the design, operation and management of the unit operations.

The 'application down' approach looks even more flawed when we consider today's needs for different qualities of water other than potable supply or municipal sewage treatment, in addition to ever more stringent quality standards. Examples might include the provision of ultra-pure water for microchip production, water suitable for crop irrigation, in-building water recycling for non-potable uses, potable supply in rural areas, water for manufacturing and treatment of industrial effluents, particularly when high in inorganic content. Very different processes may be needed to provide solutions in these situations.

SERIES MODULES

- Process science and engineering
- Principles of water and wastewater treatment processes
- Physical processes
- Chemical processes
- Biological processes
- Membrane technology
- Sludge treatment, management and utilisation
- Risk management for water and wastewater utilities

PROCESS SCIENCE AND UNDERSTANDING

supports a systems approach to utility management that includes issues such as risk management.

UNIT OPERATIONS

Understanding water and wastewater treatment through a chemical engineering approach.

When designing, operating or managing a process to provide a certain of quality water, it is better to consider first the individual unit operations that, when linked together, form the required process flowsheet for the application. This is a chemical engineering approach, as it is about the conception, development and exploitation of processes and their products. The process could be water or effluent treatment; the product a less polluted aqueous stream.

Chemical engineers use physical, chemical and biological sciences and mathematics to provide a systems approach to the understanding of changes that take place in processes, from the molecular to global scale, and to establish methods which can be employed to achieve required changes in composition, energy content, structure or physical state. In other words, chemical engineering uses a 'science up' approach to solving process problems.

This series of modules teaches the removal of pollutants from water based on a 'science up', chemical engineering approach. To understand a unit operation, the fundamental biology, chemistry or physics underlying that unit operation needs to be properly understood. Nearly all unit operations are applicable to different levels of water and wastewater treatment, e.g. ultrafiltration membrane processes can be used in flowsheets for treating municipal and industrial effluents, as well as potable supply and pure water. The fundamental concepts of transmembrane pressure, flux rates, fouling etc, remain the same, whatever the application.

The first module, *Process Science and Engineering*, covers the chemistry, biology and chemical engineering required to understand the unit operations covered in the main technical modules. Accompanying this, *Principles of Water and Wastewater Treatment Processes* introduces unit operations and offers an introduction to the remaining modules that describe unit operations by reference to their scientific principles: *Physical Processes, Chemical Processes, Biological Processes, Membrane Technology* and *Sludge Treatment, Management and Utilisation.*

This module, *Risk Management for Water and Wastewater Utilities* provides the first of the cross-cutting subjects covered in the series. The emphasis here is on techniques and approaches to risk-based decision-making within a modern operational and regulatory context.

CRANFIELD UNIVERSITY AND CRANFIELD WATER SCIENCE INSTITUTE

Cranfield University is world-leading in its contribution to global innovation. With our emphasis on the aerospace, agrifood, defence and security, environmental technology, leadership and management, manufacturing and transport sectors, we have changed the way society thinks, works and learns. We generate and transform knowledge, translating it to the benefit of society; and our partners, from micro SMEs to large blue-chip multinationals, from governments to NGOs and charities, tell us this is what they value about Cranfield. In 2015, we won a Queens Anniversary Award for Further and Higher Education for our education and research in water and sanitation for developing countries.

Cranfield Water Science Institute (CWSI) has an international reputation for its transformational research and teaching in the science, engineering and management of water in the municipal, industrial and natural environments. We have been working in water for over 40 years. Our academic and research staff, including scientists, engineers, technologists, policy specialists and social scientists are engaged in delivering postgraduate teaching, research, consultancy and training in an international arena. We run MSc programmes, support a large cohort of industry-funded PhD research students, and work closely with a range of clients across industry and government helping them address their water challenges and move their businesses towards a more sustainable green economy.

Without appropriate management, whether for municipal supply, agriculture, industry, community development or maintenance of a sustainable environment, we risk over-exploitation and contamination of our planet's most precious resource. Our core research activities are therefore focused on seven main areas:

- Sewage works of the future
- Maintaining the flow
- Water and sanitation in low-income countries
- Water for food in a changing world
- Membrane processes
- Catchment management
- Governance and asset management.

Our ethos is to provide an environment which is intellectually challenging, where research excellence is encouraged and where students, staff and ideas flourish.

Editors

Series editor, Professor Tom Stephenson

Professor Tom Stephenson leads Cranfield's research and innovation activities that include exploitation of research and development of strategy in his role as Pro-Vice-Chancellor Research & Innovation. He holds the Lorch Chair in Water Sciences since 1994. Tom is a Fellow of the Royal Academy of Engineering, a Chartered Chemical Engineer, Fellow of the Institution of Chemical Engineers (IChemE) and the Chartered Institution of Water and Environmental Management (CIWEM). Professor Stephenson is a graduate in biochemistry from the University of York and has a PhD in civil engineering, specialising in public health engineering, from Imperial College. Prior to joining Cranfield in 1990, he was a Senior Lecturer in Biochemical Engineering at Teesside Polytechnic. Professor Stephenson is also Chairman of Water Innovate, a Cranfield spin-out company which he set up. Between 2010 and 2012 he was Chairman of British Water, the trade association representing 180 water sector supply chain companies. Professor Stephenson's personal reputation for internationally leading research in water and wastewater engineering is demonstrated through publication of over 200 papers and holding a UK Engineering and Physical Sciences Research Council (EPSRC) 'Platform' grant for 10 years.

Volume editor, Professor Simon Pollard

At Cranfield since 2002, **Professor Simon Pollard** has been instrumental in growing Cranfield's environmental technology capability into a prestigious force, funding research centres in integrated waste management, risk science and environmental futures. The School of Energy, Environment and Agrifood, which Simon leads as Pro-Vice-Chancellor, represents a broad institutional capability for the resource security and low carbon agendas. Trained at Imperial College, Simon has held appointments in academe, consultancy and in Government, progressing the themes of pollution control, resource management and risk governance. He is an IWA Fellow, a Chartered Engineer, Chemist and Environmentalist, a Fellow of the Royal Society of Chemistry, and of the Chartered Institution of Water and Environmental Management. His research has been funded by EPSRC, NERC, ESRC and BBSRC in the UK and by the IWA and US Water Research Foundation, influencing the practice of risk policy within Government, among regulatory agencies and within the international water sector. He has twice been an author of Government guidelines on environmental risk management and has produced best practice guidance for the water sector on good risk governance.

Acknowledgements

This 2nd edition has benefited from feedback from numerous students since the 2008 imprint. The editors acknowledge the original contributions of current and former Cranfield University colleagues, including Prof. John Strutt, John Barnes, Prof. Michael Todinov, Dr Paul Hamilton, Dennis Bowden, Dr Brian MacGillivray, Dr Jeff Charrois, Dr Roland Bradshaw, Dr Craig Mauelshagen and Dr Ana Luís. Sections on abstraction and drought were originally adapted from lectures provided by the Environment Agency (Dr Linda Pope), and Dr Paul Gale (Animal and Plant Health Agency) provided the original supporting text on microbiological risk analysis. We also acknowledge Amanda Egerton (Egerton Consulting Limited) for her input to aspects of process risk analysis. Finally, we thank Rachel Pollard for technical editing and Maggie Smith (IWA) for editing the main text.

How to use this book

KEY POINTS

As you work through each section you will find, in the margins, key points highlighted in a box

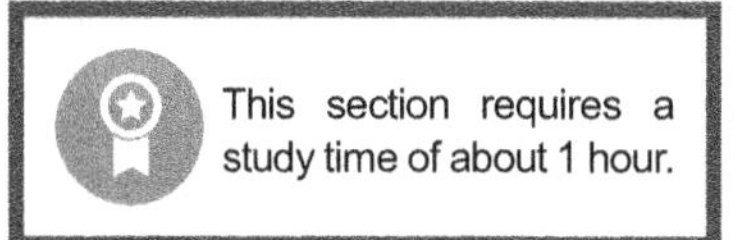

This text, or module, has been designed for *individual self-paced study* i.e. you can learn in your own time without requiring additional face-to face tuition.

Each section of the text provides step-by-step learning in sections termed *units*. Aims and objectives for each unit are provided at the start. Following these, essential prerequisites are highlighted, i.e. which other units contain concepts that you should understand before starting on the particular unit. At the start of each Section, in the right hand margin you will find an rosette symbol with an approximation of how long you will need to spend on that Section. This is an approximation and will depend on your prior knowledge and how easily you understand the subject.

The main text of each unit is enhanced with figures and tables to aid understanding. As you work through each section you will find, in the right hand margin, key points highlighted in boxes and occasional exercises to work through that help you develop your understanding of the points in the main text alongside. You will also find that certain key themes are reinforced throughout the text as you progress through units 1 to 10. In the margin there are *examples* to help understand the material in the main text. Supporting reading material id provided at the end of each unit.

EXAMPLE 1.1

Where you see the toolbox symbol, there will be exercises and examples to help you understand the material in the main text

At the end of the main text for each unit there are *self-assessment questions* that you should attempt to test your understanding. The answers, with working, are given at the end of the text. It is important that you do attempt the self-assessment questions *before* you look at the answers. This will help your understanding of the subject.

Each unit also has an accompanying set of key references.

Risk management for water and wastewater utilities

THE BASIC BUSINESS ASSUMPTION OF A WATER UTILITY

Management strategists refer to the 'basic assumption' of a business – the fundamental strategic reason that an organisation exists. The primary purpose of water and wastewater treatment is the protection of public and environmental health. This is delivered through the assessment and management of risk and opportunity.

Managing risk has become a core competency for utility managers. The provision of safe drinking water, and the protection of public health and the environment through the effective treatment of wastewaters is now routinely informed by risk-based decision-making. Aspects of utility management such as process design and optimisation, asset management and compliance monitoring each rely on a mature understanding of process risk within a broader context of business and environmental risk management. Since the first edition of this text, many utilities have created risk manager roles, implemented formalised processes for business risk management and established Board level accountabilities for the oversight of business risk. This text is a primer on these aspects of water utility governance, updated (2016) to reflect good practice and recent developments.

Implementation of the revised (2004 and 2011) World Health Organisation (WHO) guidelines on drinking water, international developments in risk-based regulation and increasing customer expectations of water quality have heightened the need for utility managers and process operators to be conversant in the principles and practice of risk management.

As part of a move towards a more strategic, forward looking approach to utility management, the International Water Association (IWA) is promoting a risk-based approach to water utility management from catchment to tap, through implementation of the Bonn Charter (2004) and recent revisions to the WHO guidelines on drinking water. Originally published in 2004, the Bonn Charter recognised the pivotal role played by water suppliers in managing drinking water quality and its publication supported their primary role in the development and implementation of drinking water safety plans (WSPs). Water safety plans have since developed into a comprehensive, preventative risk management approach from catchment to consumer, with the aim of consistently ensuring the safety and acceptability of a drinking-water supply. The IWA has been at the very forefront of their development and implementation in developed and less-developed countries.

This text is divided into the following units:

(1) Why manage risk?
(2) Basic probability and statistics
(3) Process risk and reliability analysis
(4) Assessing risks beyond the unit process boundary
(5) Regulating water utility risks
(6) Business risk management for water and wastewater utilities
(7) Managing opportunities, reputations and emergencies
(8) Embedding better decision-making within utilities
(9) Summary
(10) Self assessment question answers

Credit: istockphoto © Artfully79

Having provided rationale for the importance of risk management, the text begins with the familiar territory of unit processes and process reliability. It then broadens out to consider, first environmental risk then organisational risk management. The final sections are concerned with better utility decision-making and the development of a mature risk management culture within a utility.

Every attempt has been made to ensure that the information in this text is correct. However, should you find any errors, please let us know at: s.pollard@cranfield.ac.uk

Unit 1

Why manage risk?

COMMITMENTS TO SAFE DRINKING WATER

The provision of safe drinking water and the treatment of wastewaters and process residues are increasingly viewed in risk management terms. The 'Bonn Charter' (2004) is a strategic commitment to providing good, safe and affordable drinking water that has the trust of customers. The Charter was built on key developments in drinking water risk management led by the Australian Government and World Health Organisation (WHO). It has wide application across the water and wastewater utility sector and has influenced the revised WHO guidelines on drinking water quality (2011). These stress the importance of 'water safety plans' (WSPs) as a key process for preventative risk management.

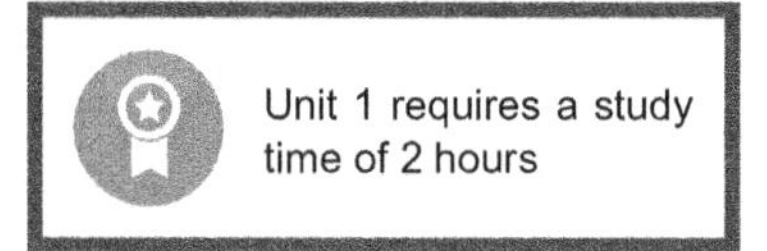

1.1 RISK MANAGEMENT AND THE WATER UTILITY SECTOR

Risk management is the primary purpose of the water supply and sewerage business. The production and supply of safe, wholesome drinking water that has the trust of customers and the treatment of wastewaters to acceptable levels of public and environmental risk are an essential societal service for affluent nations and remain a top priority for developing countries and international development bodies. Service providers, the water and wastewater utilities that, with others, manage these risks from catchment to tap and then back to the catchment face an array of complex threats and opportunities that are increasingly described in risk terms. The aim of this introductory unit is to:

- state the central importance of risk management to the overall management of water and wastewater utilities;
- provide a justification for considering risk management by reference to unit treatment processes;
- introduce the terminology, language and fundamentals of risk-based decision-making;
- describe the broader context of business and environmental risk management for utilities in which process risk analysis occurs; and
- set the study context for unit 2 on Basic Statistics and Probability.

1.1.1 Water and wastewater utilities

Utility organisations provide services to society such as the provision of safe drinking water, the collection and treatment of wastes and wastewaters and the provision of electricity to industry and homes. Historically in state ownership, many public utilities have become private sector organisations or corporatized entities, whilst others have remained in full public ownership. Whichever business model they adopt, utilities are under increasing public and regulatory scrutiny, manage vast networks of, sometimes, ageing and interconnected infrastructures and they operate increasingly in a global and multi-utility environment. Their business has become more complex; capital investment is harder and often more costly to come by; and expectations of utility performance are at an all-time high. In this business climate, utility decisions must be defensible, based on good evidence and targeted on the issues that matter most. To deliver this, utilities have to be able to competently assess, prioritise and manage risk. They must be able to put in place the organisational resources to do this and consider risks in the immediate, mid and long term.

For example, in the past the UK water industry was dominated by relatively small, reasonably 'low tech' public sector organisations with manpower costs typically in excess of half the total operating expenditure. Now, many water supply companies are either quoted on the stock exchange or form part of much larger international organisations. As a result, they are increasingly aware of the risks of major events that may impact on the future well-being of their business as well as their customer base. Sophisticated water utilities have put in place risk management structures allowing the free flow of risk information up and down their organisational structures, often coordinated by a formal 'risk manager' role, and have internal audit functions to report independently to a Board audit committee. 'Best in class utilities' have developed effective reporting structures and retained informality within operations to ensure risks and opportunities are rapidly identified and escalated through layers of line management for prioritised action.

GOOD RISK GOVERNANCE CREATES ORGANISATIONAL VALUE

Modern organisations are expected to demonstrate their capabilities in risk management. A changing business climate for utilities has focused the mind of their Boards and Executives on to good corporate risk management. Utilities that govern risks and opportunities well – that is, they can analyse, manage and communicate risks and opportunities, and put the human and organisational resources in place to do so – secure the trust of employees, customers and a wide range of external bodies.

1.1.2 Risk and regulation

The public scrutiny of utilities usually comes in the form of 'regulation' administered by government bodies to allow operations (water treatment works; wastewater works; biosolids spreading to land) to go ahead, through the issuing of permits with certain conditions attached. Taken as a whole, these 'licenses to operate' act as the 'driving licences' for utilities, allowing them planning permission and environmental permissions to abstract water from the environment and discharge treated wastewaters back to rivers, for example. Over the last 40 years, since our understanding of environmental impacts has developed, environmental regulation has become increasingly complex and wide ranging as we seek to protect public health and the environment from releases from industrial processes and other activities within water catchments.

Regulatory bodies now set challenging standards and can monitor an entire water organisation's performance against targets. The ultimate sanction on a utility is the loss of their overarching operating licence, which most believe would be a catastrophic outcome. Companies have also been taken to court by regulators for failing to meet standards or for institutional failures. Governments maintain a close eye on the 'value-for-money' that utilities offer society and, against the background of rising standards, there is considerable pressure by governments to restrict price rises to the customer. Managing the business risks to (and from) a utility has become the principal role for a utility manager. But s/he must also consider these risks in light of opportunities for growth, technological and business process innovation, diversification and business efficiency. Managing this balance of risk and reward is challenging, particularly in a sector that has historically been understandably risk averse given its public health mandate.

REGULATORY INCENTIVES

The impact of financial regulation, changes to water quality and environmental legislation, higher customer expectations, and the use of severe penalties for failing to meet standards has substantially increased the risk to water and wastewater utilities.

In response to the burden of progressive regulation, some countries are considering afresh how best to regulate utilities. 'Lighter touch' enforcement is being considered for some high-performing utilities. Utilities that can demonstrate the proactive management of their business risks are well-placed to benefit from these new types of regulatory relationship.

Regulators have to regulate an increasing number and diversity of process operations and environmental activities under their jurisdiction. Most regulators have public health and environmental protection responsibilities for a wide range of operations and activities such as nuclear power plants, integrated refineries, waste management facilities, land use, farming practices and contaminated sites alongside water and wastewater treatment facilities. Often with their own fixed resources, they must decide where best to place their regulatory efforts and, within these categories of facility, where best to target regulatory attention. Regulation itself is increasingly a risk-based activity, whereby regulators formally assess risk and then focus on activities that are likely to pose the greatest harm. Harm to public health or the environment can be posed by hazardous agents (e.g. pathogens), hazardous processes (e.g. the storage of bulk chlorine gas) and hazardous behaviour (e.g. poor operational culture and a lack of management oversight). In this sense, utilities now consider the technological, organisational and human contributors to risk management, rather than engineering risk and reliability alone.

Many regulators require utilities to demonstrate the management of operational risk – to show to them that they are capable of assessing and prioritising the management of the key threats to their business, the environment or public health. Regulators, operators and their professional advisors use risk assessments to inform decisions on operational safety and risk. They consider the magnitude and characteristics of risk alongside the costs of risk management, relevant social issues, the availability and capacity of technologies and the suitability of management arrangements required to manage key risks. Considered in this light, risk management is truly a collective enterprise between the regulator and regulated community.

LIMITATIONS OF RISK ASSESSMENT WITHOUT MANAGEMENT ACTION

Risk assessments are used by utilities and regulators to inform decisions on risk management. These decisions also require consideration of cost, social issues, technology and management.

People manage risk, and so risk assessments themselves cannot guarantee safety or protection. Risk assessments are used to inform a set of priorities for managing unacceptable risks that, in turn, result in actions that individuals take responsibility for and act on.

1.2 SHOULD ORGANISATIONS MANAGE RISK AND OPPORTUNITY?

The answer to this is a resounding 'yes'. At their core, utilities perform a public health service for society by supply drinking water and dramatically reducing our exposure to harmful or undesirable pathogens and chemicals (through water treatment, waste management, wastewater treatment, electricity generation and supply). Without effective risk management, an adverse, preventable event could pose unacceptable risks to individuals, with the additional consequences that the company could lose its reputation and ultimately its business. All businesses involve risk, if only financial risk. Organisations that have to manage financial risks, health and safety risks and environmental risks have a more complex task to perform.

Utilities must also seek and manage business opportunities. In an era of fast-moving technology and organisational change, they must consider if, how and when to introduce innovative technology to improve treatment and save energy and chemicals for example; the implementation of new data and information technologies; new workplace practices; outsourcing models; and staffing strategies. A clear tension for water utilities is how to respond to these opportunities without compromising their organisational performance, the costs of service or effectiveness of their risk management efforts.

Water and wastewater utilities that have become highly proficient at managing risk and reward are beginning to see this capability as an asset in its own right. Good risk governance creates value for organisations – it creates confidence among others, builds trust and reputational credit among customers and can have a direct impact on reducing the 'cost of risk' through lower insurance premiums, for example.

INSTITUTIONAL AUDITS AND RISK

Trends in the auditing of company accounts are providing closer scrutiny of organisations' abilities to assess and manage business risk. Most utilities will have to publish a public statement of how they manage their business risks, as part of their annual accounts, and some may list their top risks and the measures they have in place for their active management. Codes of corporate governance, especially since the economic crisis of 2008, are placing ever-greater emphasis on organisations to make their risk management procedures transparent.

1.2.1 Do organisations manage risk and opportunity?

Even in a business climate where so much emphasis have been place on risk management, some organisations are still managing their risks without formally recognising it as risk management and, in cases, without using formalised risk management processes. This was the case for many business sectors up until the 1990s. This is changing, however, with the advent of new accounting and business auditing standards, including the international standard on risk management ISO

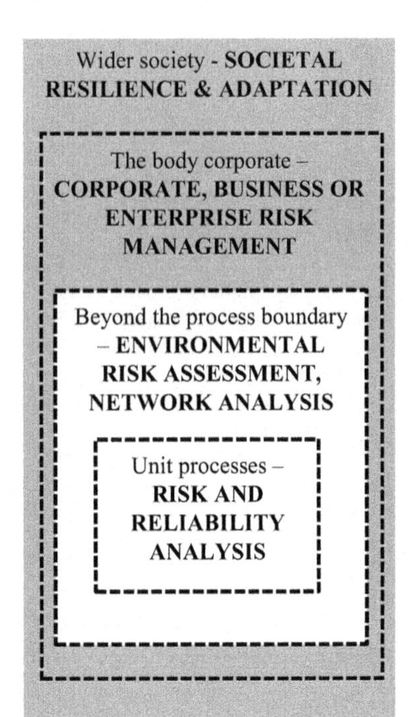

31000: 2009 *'Risk management principles and guidelines'*. This international standard, developed following some 15 years' experience with a related Australian/ New Zealand standard on risk management, provides basic principles, a guiding framework and a formalized process for managing risk. It is applicable to any organization regardless of size, activity or sector and has been widely implemented in the water utility sector. Its implementation can help organizations deliver their corporate objectives, improve the identification of opportunities and threats and allocate and use resources for risk management effectively. In many countries, public and private sector organisations are expected to assess, public declare and demonstrate how they are managing their key business risks.

1.2.2 Do organisations manage their risks and opportunities well?

Whilst most organisations couldn't stay in business unless they operated in a safe and responsible manner, if one was to consider their business as a whole and the full range of hazards to which they are exposed, in many cases the answer to this question might be 'no'. There are many reasons for poor or inadequate performance: lack of knowledge and an incorrect perception of risk, an inadequate allocation of resources, the inadequate provision of risk assessment tools, historical practices, poor competence, poor leadership, inappropriate prioritisation of organisational goals. Clearly the more complex the organisation, the greater the scope and scale of risk and therefore, the more important it is that companies employ capable risk managers. Growing awareness of the need to improve risk management, together with a series of international initiatives and the availability of guidance on risk management for the water sector, is improving this situation. Since 2004 and publication of the Bonn Charter, progress in the sector has been rapid and many utilities have recognised the need to make their implicit commitment to risk management explicit in the form of a renewed organisational commitment, the development of procedures and frameworks and Board/council level commitments to the improved oversight of risk.

This call to action has been amplified by a nagging prevalence of water quality incidents that continue to occur in the sector. Risk management for water and wastewater utilities is complicated by the nature of water supply and wastewater discharge. Unlike many sectors, there are no opportunities for product 're-call' when supplying drinking water to customers, or discharging treated wastewaters to the environment. Hence service delivery, ideally, has to be right first time, every time. Often, by the time that evidence emerges that a failure has occurred, it is too late to intervene and prevent exposure to a hazard, simply because the water has already been supplied or the wastewaters discharged. Better realisation of this very limited reaction time is forcing a preventative approach to risk management in the water and wastewater sector; that is, utility managers must now anticipate and secure early warning of the systemic changes in catchments, treatment and distribution systems that might result in an adverse incident and put in place measures to prevent incidents occurring.

In this second edition of *'Risk management for water and wastewater utilities'*, we are also concerned with risk as it relates to aspects of organisational foresight (the capacity to anticipate risks and opportunities by looking to emerging trends in the long-term), resilience (an ability to recover from a disruption to service), robustness (the property of a structure, process or organisation that represents an ability to withstand shock) and adaptation (an agility to manage change and recognise multiple pathways to a new system state). These terms have become important in the water sector as we have recognised the increasing prominence of certain drivers of change, notably climate change, land use change, demographic change and the deterioration and interconnectedness of our infrastructures. Risk and opportunity management in this complex landscape of issues in a true, preventative sense is rapidly becoming a core competency of utility managers.

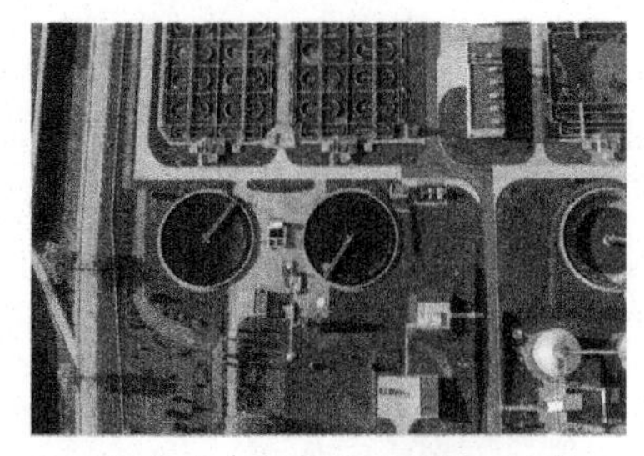

Credit: istockphoto © MariuszSzczygiel

1.3 THE ORIGINS OF RISK

An important starting point for discussing risk management is to understand what is meant by the word 'risk'. The discipline and science of risk has its origins in gambling, using games where there is an element of chance, or uncertainty, over an outcome and therefore an opportunity for gain or loss – these are the key elements of risk and of business.

RISK AND OPPORTUNITY

There is merit considering for a moment the importance of risk and opportunity in the water sector. Innovation requires us to take an element of risk in new processes, for example. When we innovate treatment processes we seek to manage the likelihood of adverse outcomes by undertaking research and pilot studies so that when we 'scale-up' process innovations, they are less likely to fail during operation. In contrast, there are some risks we manage very tightly, respectful of the consequences that might be realised were they not to be well-managed; the risk of filter 'breakthrough', for example.

For risk managers in the water and wastewater utility sector then, they must be both open to opportunities that can be managed well and to the need for preventative risk management for potent hazards. Understanding the relative character of these risks, how they may manifest themselves in water utility and practical mechanisms for their management is an expert skill.

1.3.1 A familiar example of risk

Consider a simple gambling game that involves tossing a coin. At the toss of a two-sided coin, (the initiating event), there are only two outcomes ('heads' or 'tails'). You can only either win or lose (the consequences of the outcome). If the coin is a fair coin, the probability of a head is 1 in 2, or 0.5 and the probability of a tail is also 0.5. A typical gamble might be between two people who stake, say, £1 each (in which case there is £2 in the kitty). The game requires an agreed rule (Table 1.1). For example: if heads lands face up, I win; if tails, you win. The coin is tossed and the winner takes £2 and the loser loses her stake. The first time you play this game you either win or lose. If you keep on playing this game how much do you win?

Table 1.1 Winnings in a simple game with defined outcomes.

	Winnings		Cumulative Winnings		Net Winnings	
	A	B	Cum A	Cum B	Net A	Net B
toss 1	2	0	2	0	1	–1
toss 2	0	2	2	2	0	0
toss 3	2	0	4	2	1	–1
toss 4	0	2	4	4	0	0
toss 5	0	2	4	6	–1	1
toss 6	0	2	4	8	–2	2
toss 7	2	0	6	8	–1	1
toss 8	2	0	8	8	0	0
toss 9	2	0	10	8	1	–1
toss 10	0	2	10	10	0	0

In this familiar example, the odds = probability of winning/probability of losing = 0.5/0.5 = 1. So, for a £1 stake, you make £1, (odds x stake) + your original stake. On average you win as much as you lose in this game. Although you may well be lucky and have a run of heads, over the long term you will always break even.

The key difference between games of chance, such as a lottery, and a business is that the odds can be pre-set in a lottery. For business managers, they determine the odds of 'winning' or 'losing' by the way in which they manage their business and the choices you make, for example over technology, asset replacement or maintenance, their choice of suppliers and contractors and how they are managed, the approach to regulatory compliance, the training of staff and the 'tone at the top' or priorities given to essential aspects of public health, finance and customer engagement.

1.3.2 Business risk for water and wastewater utilities

Although a business is far more complex than a coin-tossing game, it is fundamentally similar. An investment (a stake) is made in a water treatment plant, say, which is built, operated and maintained over the length of its service life to deliver a safe, reliable and trustworthy product (drinking water) that generates revenue from the tariff. The initial investment is made in the expectation that it will return a financial surplus or profit that can form a basis for future investment in the business, say in the asset base. The odds must be favourable in the long run if investment is to be attracted and maintained at a reasonable cost. That is, there must be the goal of making a surplus, even if this is returned to the public purse for reallocation the following year. However, there is still a gamble involved because the operating costs may exceed the revenue, for example. A typical business model is illustrated in Figure 1.1.

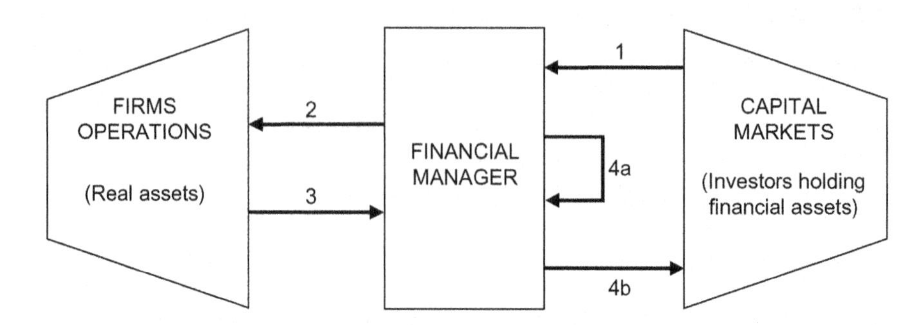

Figure 1.1 A typical private sector utility business model.

In this model, investment cash is raised from the capital markets (1) e.g. as shares, bonds or a bank loan and these funds are invested in the firms operations (2), (e.g. a water treatment plant). These are treated as costs and include all capital and operational expenditure. The operations of the firm develop revenue (or income; 3). If the revenue exceeds expenditure (the difference being the surplus or profit), it is returned to the investors (4b) and/or reinvested into the company (4a). The latter is acceptable if it leads to an increase in the value of the share price.

Business managers deal with the cash flow. Their goal is to minimise expenditure and maximise income. Activities that increase expenditure are scrutinised. If expenditure exceeds income then the company will be at risk of going out of business or at risk of takeover by a more efficient management.

This very basic model of business exposes a set of essential tensions in the operation of any organisation, whether in public or private-sector ownership. One can imagine the different motivations of key players in the model as it applies to water utilities and also the essential need to regulate the system so that a creative tension of motivations is secured to prevent so-called market failure of those operating within it. For example, engineering assets delivering safe drinking water that customers pay for need to be maintained and occasionally rebuilt so capital and operational expenditure is essential; investors expect a return on their investment; financial managers must tensions these requirements keeping both content.

Where does risk come into this business model? Unfortunately, neither income nor expenditure is certain. Events occur that reduce income and cause increased expenditure. For example, mains bursts causing leaks lead to increased expenditure on leak detection and pipe repair, refurbishment or replacement. Or an incident may occur that requires an emergency response and this will also incur operational costs and possibly compensation for customers. Techniques of risk assessment and management are applied for at least one, and usually several of the following reasons:

Financial

Both capital, and operating costs can be reduced by carrying out assessments during the design stage and project management of a scheme, or operating and maintenance costs can be optimised on an existing plant. Failures and their consequences for the business can be expensive to mitigate and can in some industries lead to the loss of the customer base. Frequently whole programmes of asset management are underway in a utility that require the input of expert project, or programme managers whose role it is to manage the uncertainties across a portfolio of projects, switching resources between them, bringing some project deadlines forward and relaxing others as the practical realities of managing complex capital build projects kick in. An international publically available standard, PAS 55, covering the management of physical assets is available and widely used among water and wastewater utilities to bring some rigour to the management of these types of multi-project activity and the project and programme level risks involved.

Impact on the business

For most companies, the biggest fear is the loss of the customer base with reduced sales of the product. A serious accident can result in a loss of confidence among the owners of the organisation which can drive down share prices, reputational value or customer trust, in publicly quoted companies and can persuade governing boards to make changes to the management and organisation. Following an accident, management has to put a major part of the business's effort into restoring equipment, stocks and customer confidence. Funds earmarked for other projects (contingencies) may have to be diverted in response to the incident, depleting reserves. Significant, adverse operational incidents frequently escalate very quickly to have financial impact on organisations and the advent of social media has accelerated the extent to which impacts can be felt.

Impact on people

The failure of a piece of equipment in a water supply system, or the wrong action by an employee can result in either the customers receiving no water supplies or receiving contaminated drinking water that may be injurious to health. This can affect a large number of people simultaneously with little or no opportunity for intervention until after exposure has occurred.

Management of risks to public health are generally of greatest concern for water and wastewater utilities and they attract significant attention. The safe drinking water agenda seeks to ensure these risks remain uppermost in the corporate mind set during a period of substantial sectoral change when a myriad of other business risks might detract from the core purpose of a water utility.

Impact on the environment

Equipment failure, lapses in organisational attention, or human error can lead to adverse consequences for the environment that may include excessive discharges to the atmosphere, soil/sediment, biota, or water environment. These may occur directly, or as a result of actions to mitigate the results of the failure, such as discharges of polluted water from a holding tank.

Customer care

If an organisation has target levels of customer care, considerable time and effort can be taken up by employees handling complaints and explaining to customers, and the press, what has happened. If compensation has to be paid when certain standards are not met, then it is usually far more economical and proper to reduce the risk of a failure in the first place rather than pay compensation to a large number of people (distribution of this compensation can itself be expensive). With increasing attention on the role of customers, of consumer bodies, the closer engagement of customers in utility business decisions and of value for money, the importance of transparent, preventative risk management is greater than has ever been the case.

Legislation

A country's legislation sets minimum standards for the product being produced, the handling of the constituent chemicals, the discharge of the waste products and the health and safety of the operational staff and the people living nearby. Even when there is no legislation covering some aspects there can be claims of negligence against the operating company if adverse events occur. Modern regulation seeks to encourage companies, including public and private utilities to go 'beyond compliance', to proactively demonstrate a maturity in their ability to govern risks and opportunities well and communicate this capability to investors, regulators and customers. In doing so, water utilities create value for their organisations.

1.4 DEFINITIONS OF HAZARD AND RISK

Two words mistakenly used interchangeably are hazard and risk. Hazards are hard to manage – they are an inherent feature of situations. Risks can be managed, having been assessed. **Hazard:** A situation or substance with a potential to cause harm; for example, to humans, property, the environment or a combination of these. **Risk:** The likelihood of a specified undesirable event occurring within a specified period or under specified circumstances.

Using these definitions, we can start to become familiar with the hazardous situations around us – dangling kettle leads, broken pavements and unprotected cables. These are all sources of a hazard – scolding, tripping and bruising, and electrocution. A chain of events must occur for the hazard to be realised and there must be something or someone that we value that is adversely affected by the hazard if there is exposure and the adverse consequences then occur. One conceptualisation of risk considers these components as the **source** of the hazard, the **receptor** or valued item or person at risk (e.g. the environment, an asset, children, immuno-compromised people) and the **pathway** by which the receptor becomes exposed to the source of the hazard. Without connectivity (the chain of events and the subsequent exposure) between the source, pathway and receptor, the consequences cannot occur. So immediately we can see one way to manage risk is to understand potentially harmful sequences of events, proactively intervene to make them less likely, and so preventatively prevent exposure.

ESSENTIAL FEATURES OF RISK PROBLEMS

Recognise the distinction between risk and hazard. It is possible for a large hazard to exist while exhibiting a low risk. This is because under normal conditions, management actions are taken to ensure the likelihood (or probability) of the hazard being realised is small. A large tank of pressurised chlorine gas represents a major hazard, but the risk of a release is small, if it is properly managed.

Individual risks have likelihood and a consequence. The consequences are usually a feature of the source of the hazard and the vulnerability of the thing it might affect. Whether or not the consequence comes to bear, the likelihood, depends on a chain of events that ultimately results in the thing we wish to protect being exposed to the source of the hazard.

Think about a tanker of milk, on the road, adjacent to a pristine fishing river. What is the source of the hazard and what hazard does it pose? What chain of events might happen were the tanker to 'jack-knife'. What might the consequences be?

1.4.1 A risk 'equation'

A further definition of risk views it as a mathematical combination of an event, an event likelihood (or probability) and an event consequence. It can be represented mathematically as either:

$$\text{risk} = \text{frequency} \times \text{consequences; or} \quad (1.1)$$

$$\text{risk} = \text{probability} \times \text{consequences} \quad (1.2)$$

The magnitude of risk can be illustrated on a plot of probability or frequency of occurrence versus the consequences. Consequences can be expressed in various ways but the most common are financial loss ($) or number of fatalities (N). When consequences are expressed as the log of the number of fatalities per event, this type of graph is referred to as a F/N plot, (F = log (frequency), and N = log (number of lives lost). The units of risk need to be considered if quantitative assessments are being made. For safety case studies for example, where organisations have to quantify risk to demonstrate they understand root-cause and can managed it accordingly, the term risk has the following quantitative meaning:-

Event consequence	Number of fatalities
Frequency	Number of events per year (yr^{-1})
Risk	Number of fatalities per year (yr^{-1})

If the risk under consideration is financial risk, then the units of risk are:

Event consequences	Total cost (£)
Frequency	Events per year
Risk	£/year

EXAMPLE 1.1

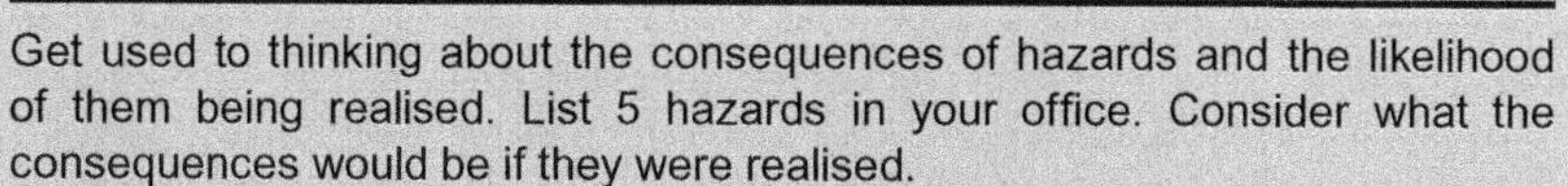

Get used to thinking about the consequences of hazards and the likelihood of them being realised. List 5 hazards in your office. Consider what the consequences would be if they were realised.

How likely are they to occur?

Considering the nature of the consequences and their likelihood, what are the greatest risks in your office? Are they being managed?

If not, is it acceptable to leave them unmanaged? What will you do about these priority risks?

Notice that for a given number of fatalities, the term risk means the same as the frequency of occurrence, and it is sometimes loosely referred to in this way. Notice also, that frequency is not identical to probability but they are related by the occurrence time distribution. The particular definition of risk adopted chosen will affect the numerical value of risk.

The risk equation can be represented diagrammatically in a plot of frequency against consequences (Figure 1.2). For example, if the consequences were to refer to the number of people made ill as a result of a water contamination, R would be the number of illnesses per event per annum, F would be the frequency of events with N people made ill and C would be the number of people made ill in the event.

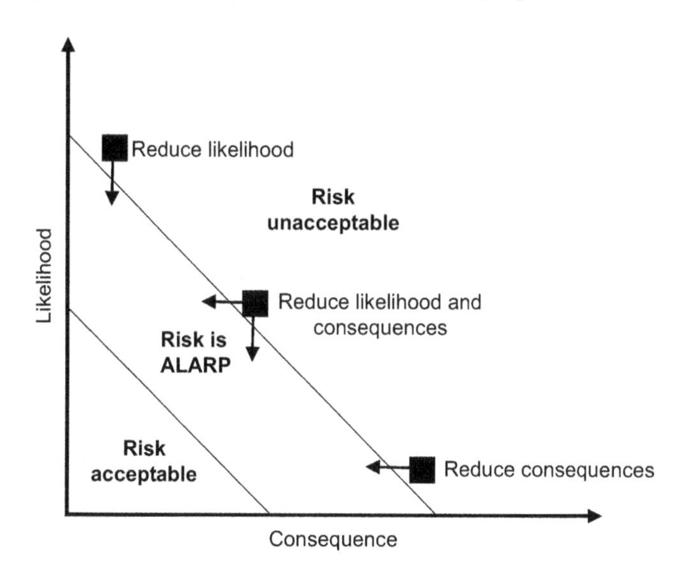

Figure 1.2 Frequency–consequence diagram.

EXAMPLE 1.2

Consider the following familiar hazardous situations:

(i) attending a public firework display;
(ii) wiring a plug;
(iii) driving in the rain.

How do manage the risk to yourself in each of these? Do you manage the source of the hazard, the pathway to it becoming realised or the receptor – you? In managing the risks, are you managing the consequences or likelihood? How do make these risk management judgements in your everyday life?

The risk plot shows three typical zones of risk: (i) one where the risk is unacceptable and requires management through a range of risk management strategies; (ii) where the risk is acceptable, having been managed down to an acceptable level; and (iii) an interim zone where the risk is managed to as low as is reasonably practicable – the so-called ALARP region. Managing a risk to ALARP infers a notion of risk tolerability and of a cost-benefit balance; that is recognition of a tolerable level of residual risk, beyond which the costs of further risk reduction are deemed disproportionate to the incremental benefit gained. Economists call this 'risk optimisation' and it suggests a point of diminishing returns has been met in terms of the financial investment in risk management measures. Discussions about where this point exists for specific hazards, and thereby the investment implications for water utilities, are contentious in the setting of water quality standards, for example.

$$R = F \times C \tag{1.3}$$

$$\log R = \log F + \log C \tag{1.4}$$

$$\log F = \log R - \log C \tag{1.5}$$

Returning to Figure 1.2, a plot of log F vs. log C has a slope of –1 for a constant risk R. A constant level of risk 1 in 1000 is equivalent to 10 in 10,000 and 100 in 100,000 and so on. On an FC plot, it is possible to compare a number of risks across a single plant or across the business as a whole, or across a number of industries. This is then called a risk 'heat map' whereby the unacceptable region is denoted in red shading, the ALARP region in amber and the acceptable risk region in green. Unacceptable become red, 'hot' risks in these schematics.

1.4.2 Risk analysis

Risk analysis is a process by which we learn about and begin to understand how accidents and incidents occur. Remember that risk analyses in themselves cannot guarantee safety and it is what you do with them that counts. Whilst some risk analyses may be quantitative and therefore generate quantified estimates of the likelihood of certain consequences occurring, it is best to consider risk analyses as diagnostic tools – they should inform where the biggest source of risk in a system is, which specific features of a multicomponent system contribute most to the risk, where the vulnerabilities are; and therefore guide where to place your risk management effort and in which order of priority. Risk analysis attempts to answer the following basic questions:

- what can go wrong?
- how might it go wrong?
- how bad could it be?
- how often could it happen?
- is that acceptable?

EXAMPLE 1.3

Risk of what to whom?

This is the first question to ask before undertaking a risk analysis – what is it we are concerned about and why.

For example, what hazards to the environment are associated with operating a wastewater treatment works?

How does responsible operation of the works minimise the likelihood of these hazards being realised?

Can you list an example of (i) an accident; (ii) a human error; (iii) an equipment failure; and (iv) an external event, any of which might initiate a chain of events that resulted in the works breaking its discharge consent?

Risk analysis has developed into discipline in its own right and there are numerous descriptions of the process with various flavours of approach. Figure 1.3

illustrates a more conventional technical process that is suitable for the start of our discussions here. Implicit to Figure 1.3 are an initial identification of the hazards; an analysis of the probability and consequences of the hazard being realised; a consideration of their combined outcome by evaluating the magnitude of the risk; some judgement of the significance of the acceptability of the risk by reference to some prior agreed acceptability criteria; and then a decision to act – either in modifying the hazard or in demonstrating that the risk is as low as is reasonably practicable (ALARP); that is the residual risk is being acceptably managed. The control loop, an iteration of the process, suggests that if actions employed first time to manage the risk are ineffectual then additional analysis and action will be required; that is, the process is iterative.

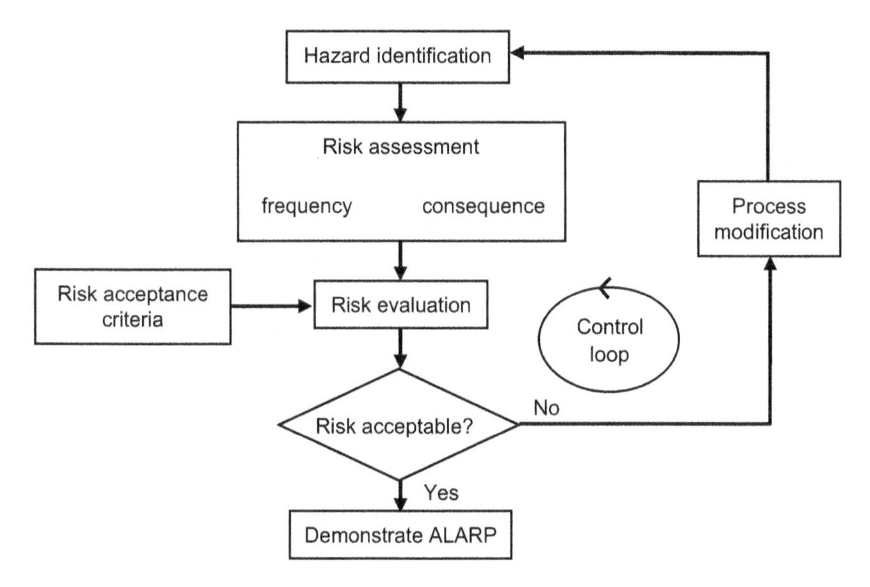

Figure 1.3 A risk analysis and management process.

Hazard scenarios may involve a number of factors including:

- accidents –e.g. road tanker crashing
- human error –e.g. switching off the wrong pump
- equipment failure –e.g. pipe burst, valve fails to open
- external hazards –e.g. flooding, tsunami, hurricane

One thing should again be made clear. Risk assessment alone does not reduce risk. Its purpose is to inform decisions on whether to, and how to manage risk and its value should be seen in that context. Management action to reduce a risk also assumes that responsibility has been passed to someone to manage the risk and that that this person, usually termed the 'risk owner', recognises themselves as accountable for action and assumes responsibility for seeing the actions through so the risk is reduced accordingly. Analysis, without action to reduce risk has limited value. Risk assessment to justify inaction, to justify the status quo, or to justify decisions already made has no credibility and will be viewed with suspicion.

TENSIONS IN PRACTICAL UTILITY RISK MANAGEMENT

There is a paradox at the heart of the water sector in that water and wastewater utilities put assets 'at risk', by the very means of operating them, in order to protect public health and the environment. Good asset management lies at the heart of managing this tension.

1.5 MANAGEMENT OF RISK

Managing risk usually means reducing risk to the organisation. There are various approaches that may be adopted including:

- risk prevention (make the situation inherently safe or reliable by design);
- risk reduction (implement physical and organisational safety barriers); and
- risk transfer (contract out the activity to others you oversee, or buy insurance).

The process modification loop in Figure 1.3 implies all of these.

1.5.1 Acceptability of risk

The results of a risk assessment lead directly to the question of what is an acceptable risk. Attempts to establish risk criteria generally follow two approaches; one has been to determine the maximum probability of an adverse consequence (e.g. a fatality) to which it is possible to expose an individual to and then judge its acceptability (the so-called statistical value of life approach, widely deployed for transport system risk decisions, say); and the second is to determine the maximum sum of money which it is reasonable to spend to avoid a fatality or particular form of harm. The first approach requires an absolute limit regardless of cost. The second is more realistic from a business management perspective but is contentious and raises emotive reactions and ethical issues. Also note that we might tolerate risks in practice whilst not finding them wholly or broadly acceptable. We tolerate the risk in light of the benefits that accrue, for example through employment (high hazard industries such as construction), personal convenience (mobile phone use) or the production of electricity (potential electrocution).

1.5.2 Risk management capabilities

Business risk management is concerned with the identification, assessment and management of all risks relevant to an organisation's operations. Risk management involves, therefore, a set of important key processes and practices that require a high degree of competence and capability. Not all organisations are equally good at managing risk. Their ability to do so, or their maturity in their capability to manage risk, will vary with the level of scientific and engineering knowledge, their skills and organisational experience, the resources employed, (time, people and tools) and their commitment. With experience, their capability in risk management will mature and this can be measured and used to drive continuous improvement. Increasingly, capability maturity models (CMMs) are used to assess organisational capabilities. This is a useful way to compare organisations, or the different business functions within organisations (benchmarking). There are two main assumptions that lie behind a capability maturity model:

RISK APPETITE INTRODUCED

Deciding on your or your organisation's appetite for risk is essential to deciding how far you will go to manage risk. For public health and environmental risks, the extent of risk reduction is usually specified in a regulatory standard, or embedded within the derivation of a water quality guideline, for example.

(i) The capability of an organisation to design and operate a plant safely is important and it is assumed that the measurement of organisational performance will yield additional information over and above that obtained from assessment of the output of specific products and processes.
(ii) There is a progression through increasing and distinct levels of maturity. Some organisations have greater capabilities to design and operate safely than others, e.g. through a combination of experience and good practices supported by research and relevant and up to date training *etc.*

In these models, capability is often ranked according to five distinct levels of maturity (Table 1.2).

Table 1.2 Levels of organisational capability.

Level	Interpretation
1	**Learner organisation.** This is the lowest level and defined by the absence of qualities linked to the higher levels. These organisations have ad hoc approaches to safety, risk and reliability management.
2	**Repetitive organisation.** At this level organisations can repeat what they have done before, but not necessarily define exactly what they do. They might repeat practices that satisfy legislation in regimes that are now obsolete, or repeat practices that only partially satisfy current legislation. These organisations have rational quality management systems and can repeatedly deliver consistent products and services. However, they do not possess defined risk management processes aimed at delivering safer, less risky and more reliable products and services.

(Continued)

Table 1.2 Levels of organisational capability (*Continued*).

Level	Interpretation
3	**Defined organisation.** At this level, organisations can define safety, risk and reliability requirements and their associated process but have limited feedback processes leading to safety improvements. Essentially, the management process is almost, but not completely open loop. These organisations understand what is needed to reduce risk and deliver higher levels of safety and reliability. They are however, limited in their ability to implement reliability improvement and risk reduction strategies in practice.
4	**Managed organisation.** At this level, organisations can control what they do in terms of design and operations to achieve high levels of safety and reliability. They can lay down requirements and, through benchmarking, ensure these are met. They are learning organisations, in that all the important characteristics are benchmarked and acted on in the feedback process, but the learning mode is single loop in that only the product or process is changed. They have well-defined organisational processes to manage safety, risk and reliability, they are limited in their ability to self-adapt and sustain their capability over the long term. There is limited organisational learning about safety, risk and reliability and relatively little education, training or R&D.
5	**Adaptive organisation.** Organisations at this level use best practice and are exemplary. They are capable of learning and adapting their organisational structures as a result of complete and effective benchmarking and organisational feedback. They practice double-loop learning; that is, they do not just use experience to correct problems and processes, they also change the way they operate. They are capable in all the key risk management processes, implement advanced safety, risk and reliability practices, and can sustain this capability in the long term.

1.5.3 Corporate risk management

Clearly, anything that changes the predicted financial outturn of an organisation will reflect on the executive Board or Directors. In the case of companies listed on a stock exchange, the share price will be affected and possibly their credit rating. For publically owned utilities their performance may impact on their future access to public funding or the degree of regulatory attention they receive. Whereas day-to-day share prices have little practical effect on a company over the long term, a lack of growth of share price will not be to the liking of shareholders. The major holders, often City institutions seek changes to the Board of Directors or the running of the company. Alternatively, the company may be a target for takeover by another company. It is the role of water utilities in the private sector to balance the provision of public and environmental health protection with good customer care, the maintenance and operation of assets and good shareholder return. Lower credit ratings result in increased interest charges on loans from financial institutions.

FEATURES OF A CORPORATE RISK MANAGEMENT STRATEGY INCLUDE:

(1) Definition and scope of the strategy, to include agreement on acceptable levels of risk and corporate risk appetite.
(2) Identification of risk, to include assessment of the consequences of failures and the likelihood of occurrence.
(3) Identification of risk management methods.
(4) Technical and economic appraisal of risk management options.
(5) Managing emergencies and business recovery.
(6) Agree a programme of action and resourcing.
(7) Monitor implementation.

Private sector companies will also receive public messages of disapproval from shareholders in the form of lower share prices. Major incidents can lead to loss of profit to private sector companies due to the response to the incident and the actions taken by the company to restore customer and shareholder confidence. In the public sector, occasionally a loss of Chairman or other senior member occurs, but usually there is a diversion of objectives of the business to prevent a reoccurrence that consumes considerable resources.

Clearly for any water utility it is essential to minimise any impacts on customer service and charges to customers. Detailed financial and business planning will minimise the inevitable changes from government or regulatory sources. In theory the local authority and government organisations are better able to withstand any major impacts, as they cannot be declared bankrupt or have their operating licence removed, unlike companies in the private sector. However, in both sectors, it is accepted by managers that all types of risk to the business should be identified, assessed and appropriate action taken to optimise them.

With respect to the operation of unit processes, which is the subject of this series of texts, we are principally concerned with risks to and from assets – the unit processes that comprise treatment plant. One conceptualisation of water treatment for example, is 'putting assets at risk' (through their prolonged operation) for the purposes of protecting public health (by supplying safe drinking water). This concept of placing assets at risk so as to protect public health is useful because it articulates the reality of utility management and distils the twin goals of, and tension between, asset and risk management. Utility managers must balance these risks in managing a safe and efficient water treatment network.

EXAMPLE 1.4

Making the financial case for risk management.

Given what you have read in this Unit on the context of preventative risk management in the water and wastewater sector, imagine yourself presenting, to a finance director or a utility Board, a case for increased investment in risk management processes, training, staff and actions.

You will be to be concise, persuasive and influential. What arguments will you use to press for this investment?

Now imagine being in a high-performing utility without a history of recent adverse incidents. What response might you expect? How might you overcome resistance?

In the next Unit, we will refresh ourselves on basic statistics and probability, some of which we will later use for rudimentary quantitative risk assessment. Keep in your mind, however, the rich landscape of risk that we have introduced here and consider the inevitable uncertainties there are likely to be an evaluating risks to and from process equipment. Given these uncertainties, expect now not to be able to fully quantify most risks with accuracy and precision.

1.6 SUMMARY AND SELF-ASSESSMENT QUESTIONS

Unit 1 has provided an introduction to risk analysis and management for the sector. The principal capability you need to develop, initially, is to start thinking with a risk mind set. You need to be able to distinguish between the likelihood and consequence components of a risk problem and identify the source, receptor and pathway contributors to risks. You might also start thinking about the technological or engineering, human and organisational features that could contribute to an initiating event or series of events that could ultimately result in an exposure. With this in place, you will be able to discern risk problems with ease, understand better how to tackle them and importantly, develop your own ideas on how best to manage risk in practice. The following self-assessment questions will help you develop this capability. Use them before progressing to Unit 2.

SAQ 1.1 Think about the game of Monopoly™. Describe the consequences of landing on 'Mayfair', with and without a hotel on it, in risk terms. Use the terms probability and consequence in your answer.

SAQ 1.2 A current concern is one of a bioterrorist attack on a water treatment works. What are the sources of a potential attack? What are the receptors? Describe the pathway – the mechanism by which such exposure may occur. Present the pathway as a sequence of numbered events.

SAQ 1.3 Potent toxic chemicals reaching an activated sludge plant through a trade effluent discharge may render the microorganisms on which treatment relies ineffective. What is at risk in this example? How would you manage this risk – would you address the source, pathway or receptor? Draw a schematic of this situation before and after risk management.

SAQ 1.4 What hazards might affect (a) an urban raw water intake; (b) an upland catchment reservoir?

SAQ 1.5 On a scale of 1 to 5, qualitatively rank the following adverse consequences of water treatment failures and justify your answer:

–an E. Coli 0157:H7 (potent human pathogen) outbreak;
–the presence of benzene in drinking water at concentrations below the drinking water standard;
–musty tasting water from the tap.

Note down the uncertainties that make your ranking difficult to complete.

SAQ 1.6 "I've heard plasticizers leach into bottled water over time and may be affecting my health. Should I be concerned?" Pen a brief answer to this customer who has called your customer helpline.

1.7 FURTHER READING

Australian Government (2004). National water quality management strategy. Australian drinking water guidelines 6, Nation Health and Medical Research Council and National Resource Management Ministerial Council, Australia, 615 pp.

Bartram J., Corrales L., Davison A., Deere D., Drury D., Gordon B., Howard G., Rinehold A. and Stevens M. (2009). Water safety plan manual: Step-by-step risk management for drinking-water suppliers. World Health Organization. Geneva, Switzerland.

Bradshaw, R., Gormely, Á., Charrois, J. W., Hrudey, S. E., Cromar, N. J., Jalba, D. and Pollard, S. J. T. (2011). Managing incidents in the water utility sector – towards high reliability? *Wat. Sci. Technol: Water Supply*, **11**(5), 631–641.

CRC for Water Quality and Treatment (2004). A guide to hazard identification and risk assessment for drinking water supplies, Research Report 11, CRC for Water Quality and Treatment, Adelaide, 115 pp.

IWA (2004). The Bonn Charter for safe drinking water, IWA Publishing, London.

Havelaar, A. H. (1994). Application of HACCP to drinking water supply. *Food Control*, **5**, 145–152.

Health and Safety Executive (2001). Reducing risks protecting people: HSE's decision-making process, HMSO, Norwich, UK. Available at: www.hse.gov.uk/risk/theory/r2p2.pdf.

Hrudey, S. E. and Hrudey, E. J. (2004). Safe drinking water: Lessons from recent outbreaks in affluent nations. IWA Publishing, London, UK.

Hrudey, S. E. and Hrudey, E. J. (2004). Ensuring safe drinking water: Learning from frontline experience with water contamination. American Water Works Association Denver, US.

Hrudey, S. E., Hrudey, E. and Pollard, S. J. T. (2006). Risk management for assuring safe drinking water. *Environment International*, **32**, 948–957.

Hutton G. and Bartram J. (2008). Global Costs of Attaining the Millennium Development Goal for Water Supply and Sanitation. *Bull World Health Organisation*, **86**, 13–19.

Lindhe, A., Rosén, L., Norberg, T., Bergstedt, O. and Petterrson, T. (2011). Cost-effectiveness analysis of risk-reduction measures to reach water safety targets. *Water Research*, **45**, 241–253.

Pollard, S. J. T., Strutt, J. E., MacGillivray, B. H., Hamilton, P. D. and Hrudey, S. E. (2004). Risk analysis and management in the water utility sector – a review of drivers, tools and techniques. *Trans. IChemE, Part B*, **82**(B6), 453–462.

Summerill, C., Smith, J., Webster, J. and Pollard, S. J. T. (2010). An international review of the challenges associated with securing 'buy-in' for water safety plans within providers of drinking water supplies. *J. Water & Health*, **8**, 387–398.

World Health Organisation (2011). Guidelines for drinking water quality, 4th edn, WHO, Geneva.

Unit 2

Basic statistics and probability

KEEP IT SIMPLE

To assess and manage risk, you must be confident in distinguishing between an initiating event, the likelihood of its occurrence, the outcomes that may rise if the event occurs and the ensuing exposure and harmful consequences.

Consequences are usually characterised in terms of harm to something we value. Use probability theory to consider the relative likelihood of an event occurring or, for an event that has already occurred, for assessing the relative likelihood of a number of subsequent outcomes with ensuing consequences.

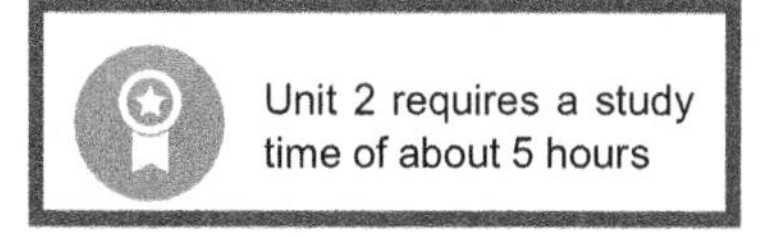
Unit 2 requires a study time of about 5 hours

2.1 INTRODUCTION

Risk analysis requires an understanding of uncertainty and probability. Even where qualitative or semi-quantitative techniques are used, being able to recognise and apply these concepts in practice is essential because understanding probability, consequence and the uncertainty of both lie at the heart of decisions about how to manage them. The distinction between the likelihood of an event occurring and the subsequent consequences of the event should it occur is a key concept in risk-based decision-making. Risk management focuses on managing either the likelihood or consequences of (usually adverse) events; and strategies for managing risk, informed by risk analysis, focus on managing uncertainty in these components of risk, prioritising risks by reference to their probabilities and consequences and addressing the most important risks first.

Many environmental and system reliability problems (unit 3) have to cope with sparse data, and so we adopt statistical techniques to assist with these limited data sets. A note of caution however. There is often a temptation to ascribe greater power to numerical expressions of risk than can be supported by one's knowledge of the system under study. Misplaced accuracy and precision in risk estimates may lead to misplaced confidence in systems. So, in setting out on a summary of statistics and probability theory in support of risk analysis, we need to be conscious of the limitations of these techniques, especially in light of the data they rely on and employ.

Statistical methods exist for dealing with data that are either in numerical form, or can be converted to this form. In general, the data available usually represent a sub-sample of some greater data set and therefore shows variability; often substantive variability in orders of magnitude. For processes such as water and wastewater treatment, several data sets may be under consideration and used concurrently, each with differing levels of variability. In order to use this data in a meaningful way, information must be obtained from the collected data, in the presence of this variability, to support decisions about the underlying data population. There are some standard statistical 'recipes' that can be applied for this, but the most benefit will usually be obtained from adopting a 'questioning' approach, based on general statistical concepts and a few well-known tools.

Before setting out, it is also worth reminding ourselves that most engineering reliability data are hard to come by. Furthermore, not all data has the same provenance, or quality of origin, so this should be interrogated first. The enquiring mindset should be focused on aspects of a problem that pose the greatest uncertainty or the greatest probability or consequences. Getting hung up on data precision is unhelpful – efforts in risk analysis are best spent in identifying key contributors to the risk, understanding their variability and the characteristics that can be managed to reduce risk. In this sense, quantitative assessments are best employed diagnostically to provide insight to the problem or system under study (e.g. the likelihood of a treatment failure; the integrity of a distribution system; asset conditions and the implications of failure). The risk analyst must keep at the front of their mind the fact that risk assessments on their own do not reduce risk – their principal utility is in informing risk management, which itself requires implementation and monitoring before reductions in risk occur.

2.2 KEY POINTS

2.2.1 Variation

In measurements of the lifetimes of physical systems such as water and wastewater unit processes (assets) and their components, variation in the results will always be observed. This variation has to be measured, and allowed for, in order to model the distribution of component lifetimes which are essential to decisions about when to preventatively maintain, refurbish or replace a physical asset.

2.2.2 Data collection and validity

Sometimes data just happen to have been collected in the absence of any defined purpose and an attempt may be made to extract something useful from them. This may be partially successful, but the important thing is that one purpose of any analysis should be to guide the type and format of data to be collected. Samples taken to provide the data should be representative of the population from which they come. There should be a sufficient element of randomness about their collection and selection to ensure the validity of any statistical techniques used and the avoidance of bias. Unfortunately, this is frequently ignored when data are used for risk analysis.

On a practical note, if you have set up a data collection procedure with a specific objective in mind, it is a good idea to be physically present, at least the first time sampling occurs. If you are not, you may obtain unsuitable data – it is surprising how easy it is for a third party to inadvertently record sample information that is incompatible with the requirements of the analysis that another user may subsequently undertake.

MANAGE YOUR RISK DATA AND RISK KNOWLEDGE

Have you considered how data will be handled, its quality checked and its validity audited? What about when systems change? Are procedures in place to ensure the data on risk analyses rely are current? Were you physically present when the data were collected? If not, what level of confidence do you have in the data and its provenance?

2.2.3 Population and sample

In many circumstances we are concerned with the failure or, more broadly, with the reliability of physical systems or component parts (e.g. valves, pumps, connections, networks). Life-testing is usually destructive, or at least degrading, and therefore it is neither sensible nor possible to test a whole population (*i.e.* all there are or could be) of systems or sub-systems to find the pattern of their component lifetimes. It can be expensive and it is therefore preferable to use a sample of test items. The number of test items in the sample should be the smallest consistent with the objective of obtaining adequate information for decision-making.

The members of a randomly selected, representative sample can be examined in detail and the findings used to make inferences about the unknown properties of the population. We are usually most interested: (i) in what the typical values of a data set are (mean cadmium content of an urban wastewater sludge; mean lifetime to failure of a pump); (ii) in the variability of one or more properties within the population (e.g. a valve lifetime); and (iii) about how any of these properties is distributed through the total population.

2.2.4 Probability explained

Probability then, is the name given to the measurement of chance, with a scale of values ranging from 0, representing impossibility, to 1, representing certainty. Most non-trivial events have an associated probability other than zero or one. For example, the reliability of a component, or a system, can be expressed as the probability that the component, or system, will perform a defined function in a specified operating environment for a given time, (e.g. the performance of a filter). A calculus of probabilities has been used over the last few hundred years but there is still considerable debate about how probabilities should be defined and interpreted.

The Bayesian school sees subjective probabilities as admissible, whereas frequentists reject this approach; the latter requiring probabilities to be obtained objectively and the only real element of subjectivity coming in how an individual will interpret a probability. In other words, what risk he or she will take of being wrong in their decision. The frequentist approach will be taken here.

Many organisational accidents that result from a single initiating event (e.g. the failure of a filter; the overtopping of a tank; the bursting of a mains pipe, the overtopping of a flood defence), have complex and deep seated root causes within systems (and organisations). Given a set of individual events that can be combined into more complex events, it is often possible to find the probability of the complex event by manipulating the probabilities of the individual linear chain, or sequence of linear events that precede it. This is applicable to linear, unit process systems.

BASIC BOOLEAN LOGIC

This describes how probabilities should be combined. For the probability of A or B, written as P(A or B) or $P(A \cup B)$ the individual probabilities are summed. For P(A and B) or $P(A \cap B)$, they are multiplied. If the probability of an event occurring is p, then the probability of that event not occurring is (1 – p). See example 2.1

EXAMPLE 2.1 WORKING WITH STATISTICAL INDEPENDENCE

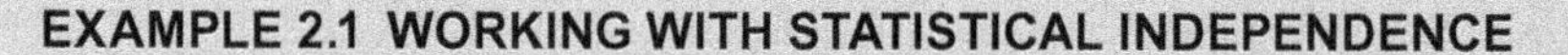

Two water pumps, an old one (A) and a new one (B), have been installed and work independently from one another. The probability of finding the old pump still in operating condition after one year's service, $P(A)$, is 0.6; for the new pump, this is 0.8. Calculate the probabilities of the following events:

(i) both pumps working at the end of the year:

$$P(A \text{ and } B) = 0.6 \times 0.8 = 0.48$$

(ii) at least one pump working at the end of the year:

$$P(A \text{ or } B) = (0.6 + 0.8) - (0.6 \times 0.8) = 0.92$$

(iii) both pumps failed at the end of the year;

$$P(\text{not } A \text{ and not } B) = (1 - 0.6) \times (1 - 0.8) = 0.08$$

(iv) at least one pump failed at the end of the year.

$$P(\text{not } A \text{ or not } B) = (1 - 0.6) + (1 - 0.8) - (1 - 0.6)(1 - 0.8) = 0.52$$

We are familiar with the use of Boolean logic in combining constituent probabilities. If an event is said to occur, if at least one of several constituent possibilities occurs, the total event probability is the sum of the constituent probabilities. For example, the probability of getting an odd number on a single roll of a die is the sum of the constituent probabilities of the faces 1, 3 or 5, (the odd numbers), on a six-sided die; this is 1/6 + 1/6 +1/6 = 1/2, a result that could have been deduced directly from our knowledge of the properties of an ideal die, (3 out of 6 'equally likely' faces each have odd numbers on them).

On other occasions, probabilities of constituent events must be multiplied together. Again using dice as a model, the probability of throwing two 'sixes' on two throws, (or on 2 dice thrown once), is given by $1/6 \times 1/6 = 1/36$. In modelling the reliability of complex systems, addition and multiplication of probabilities are the basic operations needed, although account must often be taken of the conditional nature of events and/or of their not being mutually exclusive (See Example 2.1).

Sometimes the probabilities of events occurring need to be modified by the prior occurrence of earlier events. For example, a (poorly maintained) twin-engine aircraft may have a failure probability of 0.01 for each engine on a given flight. Assuming that engine failures are independent, the probability of a 'system failure', (*i.e.* both constituent engines failing during the flight), is $0.01 \times 0.01 = 0.0001$. If one engine fails though, it is likely that the plane (the system) can still be flown on the remaining engine. However, since the surviving engine would then be more stressed, its failure probability for the remainder of the flight is likely to be higher than it would otherwise have been because of the prior failure of the other engine. Thus the *conditional* failure probability of the system would be higher than the value of 0.0001 given by the simple model based on independent events.

Implicit in this example is the notion that engine failure would only be caused by engine faults. In practice, it might also occur because of the failure of other elements such as the fuel delivery system. The overall system failure probability would therefore be increased further. It would be given by the sum of the separate failure mode probabilities modified according to whether these modes were mutually exclusive or not. Though extremely unlikely, it is possible that engine failure and fuel delivery failure could even occur at the same time and thus these two events are not mutually exclusive. For two events, A and B, that are not mutually exclusive, the following relation applies:

$$\text{Prob}(A \text{ or } B \text{ or both}) = \text{Prob}(A) + \text{Prob}(B) - \text{Prob}(A \text{ and } B) \quad (2.1)$$

If A and B **are** mutually exclusive, then they can not both occur on the same observation and the relation becomes:

$$P(A \text{ or } B) = P(A) + P(B)\ [P(A \text{ and } B) = 0] \quad (2.2)$$

Suppose the reliability of a hazard warning device (e.g. a chlorine release alarm at a water treatment plant) for a 24-hour period is known to be 0.9 (90% reliable). This can be interpreted to mean that in the long run, the system will correctly monitor potential hazard conditions on 9 days out of 10. This also means that, on an average of one day in every ten, the system will fail and that chlorine release may continue unchecked until the sensor failure is discovered and repaired. If we suppose the sensor is tested every 24 hours at noon, and that subsequently its reliability will remain at 0.9, then a 10% risk of failure at some point during any day is probably too big a risk for most utility managers to accept. Of course, even if the sensor is unknowingly out of action for some part of the day, the chance of a chlorine release occurring in this period may itself be small, so the net risk of an unchecked release would then be much smaller than 10%. Even so, most of us would not tempt fate.

The sensing system can be made more reliable by installing two sensors. If we assume the system is reliable, providing at least one of the sensors is operational, then the alarm system will fail only if both component sensors fail during the same 24 hour period. In this case, the failure probabilities of the component parts are multiplied together to give $0.1 \times 0.1 = 0.01$. The system has been made more reliable (99%), but at twice the cost because now two sensors are deployed. This is an example of making a reliable system from (relatively) unreliable components through the use of system redundancy. We have purposely introduced 'organisational slack'. In practice, the problem might more usefully be worked by first specifying the system reliability required and then finding how many component sensors would be needed to meet a specified reliability criterion. If the cost is too great, then there is a trade-off between system reliability, (*i.e.* the benefit of reliable risk management), and the cost of achieving it. At least if the kind of probability modelling outlined has been done, the risks can be quantified, even if only roughly.

For risk-critical systems, introducing this level of redundancy has clear benefits, but in cost-constrained environments there may be management pressures to remove the 'luxury' of perceived redundant alarms in the interests of organisational efficiency. One role for risk managers is to identify when these pressures are applied and challenge this rationale. More often than not, such moves are a false economy.

2.2.5 Sampling error – variability

The inferences made from a sample depend on which members of the population are included. Given the random sampling element, it is possible to make 'best' estimates of population parameters allowing for sampling uncertainty. This allowance takes the form of stating a 95% (say) confidence interval. Confidence intervals are widely used within the water and wastewater sector for analytical data; say by reference to a wastewater discharge consent from a sewage works.

The calculation of a confidence interval uses known theories of how random samples behave when selected from a defined population. The exact size of the confidence interval depends directly on the natural variability between members of the population, inversely on the square root of the sample size used and is an increasing function of the level of confidence required. So, a 95% confidence limit for a population mean is given by:

$$\bar{x} \pm 1.96\frac{\sigma}{\sqrt{n}} \qquad (2.3)$$

Where:

$\bar{x}$ = sample mean of n observations
σ = population standard deviation
1.96 is a factor related through the normal distribution to the value 95%
(For 99% confidence, the multiplying factor increases from 1.96 to 2.58.)

If, as is usual, the population standard deviation is not known, σ can be replaced by its sample estimate, s, and 1.96 is replaced by a (larger) number obtained from the t-distribution, this actual number depending on the degrees of freedom $(n - 1)$ of the estimated variance s^2.

2.2.6 Hypothesis testing

In the above section, a sample was used to estimate a likely range of values within which a related population parameter lay; say the ammonium concentration of a consented discharge sample. An alternative approach is to postulate a specific value for the population parameter and then assess whether the sample value is consistent with it. The degree of consistency is judged in terms of the probability of observing such a sample value, assuming the truth of the hypothesis. A low probability (5% or less, say) may suggest the hypothesis is not true.

This 'low' probability that guides the decision about the test hypothesis is a measure of the risk that the decision-maker is willing to take about wrongly rejecting a test hypothesis which is, in fact, true. If the consequences of such a wrong decision are very high, then a probability of much less than 5% (1%, 0.1% or 1 in a million, say) would be more appropriate. This probability, known as a Type I error, is sometimes referred to as a p-value. These p-value probabilities of a Type I error only refer to the statistical significance of the result. This may not be at all the same thing as a result of any practical significance.

For example, imagine a tossed coin has a probability of 0.501 of landing with 'heads' uppermost. If you completed a large enough number of tosses, you could demonstrate at any p-value you chose that the coin was biased in favour of 'heads' and therefore against 'tails'. Choosing a p-value of 5% and a probability of 90% that the actual bias of 0.001 would be detected would require about 2,624,400 tosses! Put another way, the behaviour of this coin when tossed a small number of times would be indistinguishable from that of a perfectly fair coin. Thus, although a statistically significant bias can be shown in principle, it would have no practical significance and the coin would be entirely suitable to use, say to decide who has first choice at the start of a sporting contest.

What about a Type II error? The Type II error would be the probability of failing to reject a test hypothesis when it is in fact false. The size of the Type II error depends not only on the p-value chosen and the sample size used, but also on the degree to which the hypothesis is false. In the coin example above, this discrepancy was $0.501 - 0.500 = 0.001$. If the 2,624,00 tosses were carried out, with a specified two-sided Type I error of 5%, then the Type II error would be 10%. This follows because we only specified a 90% chance of detecting the assumed bias of 0.001 in the fairness of the coin. Conversely, the risk of not detecting the bias of the coin is 10%. The Type II error is closely related to the practical size of difference between the hypothesised value and the true population characteristic that it is important to detect.

2.2.7 Size of sample

In a test of hypothesis, the magnitude of the two types of error and the sample size used are not independent of each other. In general, any two can be fixed and then the third will be determined by the actual characteristics of the population being sampled. Ideally, the size of the two opposing kinds of risk should be set and then the necessary sample size determined. Sometimes when this is done in practice, the sample size is unacceptably large and the trial is run with whatever size is feasible. In this case, a p-value is decided on and the Type II error is left to float at some unknown value. Whether or not this matters depends on the actual properties of the sampled population. Unfortunately, it is also quite common practice to specify the sample size and work with a 5% significance level without regard to the consequences of a Type I error or consideration for the possibility of a Type II error.

In those situations where the amount of data available is limited, care should be taken to assess the validity of conclusions bearing in mind the kinds of errors that have just been discussed.

2.3 PRESENTING STATISTICAL DATA

There are many simple statistical tools to facilitate useful analysis and interpretation. These are described below.

2.3.1 Histogram

This is a useful way of summarising a set of numbers. It indicates the average and variability of the numbers and the shape of their distribution. The larger the data set on which it is based, the more reliable these subjective estimates of the population properties become. The histogram, or bar chart, is a useful tool for communicating information to anyone within an organisation.

2.3.2 Time plot

The basic histogram destroys any time sequence in the data. A graph of the measured property against time, or a spatial characteristic say, is useful for examining step changes, trends, cyclic variations or combinations of these. The number of the initial members of a group who have survived at a given point in time is a particular example of a time plot. It is a common device used in reliability analysis.

2.3.3 Measures of average

Often, the sampled data must be summarised in numerical terms. There are various measures of so-called 'central tendency', (*i.e.* of typical or average value), available. The most common and frequently the most useful is the arithmetic mean. The median is sometimes more convenient to use. If x_i represents the ith member of a sample of size n, then the arithmetic mean, $\bar{x}$ of the sample is calculated as:

$$\bar{x} = \frac{\sum_{i=1}^{n} x_i}{n} \tag{2.4}$$

If the sample is actually the whole population, then this calculation provides the value of the population mean, μ.

The median is the mid value of the observations when they are ranked in ascending order by value, (where n is odd). If n is an even number, the median is the arithmetic mean of the two middle values. The median is a useful measure of average when a few observations show extreme deviation from the majority. It may also be useful in shortening a life-test programme by stopping the test when half of the items have failed.

2.3.4 Measures of variability

Several measures of variability have been defined but the standard deviation is most widely used. It is obtained by first calculating the variance of the data whose square root then gives the standard deviation. Variance has the useful property of additivity when considering the combination of independent sources of variation. It is this property used in partitioning a total variability into two or more component sources of variation in the technique known as analysis of variance (ANOVA). Standard deviation also has a convenient relationship with the well-known 'normal' distribution. Variance is the mean squared deviation of the sample data about their mean. In symbolic terms it is given by:

$$\text{Variance} = \frac{\sum_{i=1}^{n} (x_i - \mu)^2}{n} \tag{2.5}$$

This formula applies when the value of the population mean, μ, is known. If the values of x are those of the whole population, then the variance is given the symbol σ^2. More generally, the values of x are only a (random) sample subset of the population and the value of μ is not known. In this case, in order to obtain an unbiased estimate of σ^2 the formula used is:

$$s^2 = \frac{\sum_{i=1}^{n} (x_i - \bar{x})^2}{n - 1} \tag{2.6}$$

The divisor, $n - 1$, is equal to the degrees of freedom of the numerator. This follows because there are only $n - 1$ independent pieces of information in the numerator because of the constraint that:

$$\sum (x - \bar{x}) = 0 \tag{2.7}$$

An estimate of standard deviation is given by

$$s = \sqrt{\frac{\Sigma(x - \bar{x})^2}{n - 1}} \tag{2.8}$$

This expression is the one usually adopted, especially with small samples of less than about 20 or 30. The reason for this is that the population mean, μ, is rarely known and its sample estimate, $\bar{x}$, has to be substituted. Most calculators have a facility to use either the n or the $n - 1$ divisor. For the repeated use of very small samples, (at most, about 8), the sample range is frequently used as a measure of variability, especially in statistical process control applications.

2.3.5 Probability distributions

If a characteristic is measured for each member of a population, the observations will usually vary. A histogram of these data will indicate how the measurements are located over the range of the data. It rarely happens in practice that every member of a population is measured; a sample only is usually available. It is assumed, at least where random, representative samples are taken, that the shape and extent of the histogram will tend to be more like that of the whole population as the sample size is increased. However, with small samples, considerable sampling fluctuations can be expected.

It is convenient to represent the shape and numerical properties of a population in terms of a mathematical description known as a probability distribution. Assuming that a population is well described by such a function, this function can be used to make probability statements about the properties of future sample members taken from the population.

2.3.6 Discrete probability distributions

There are many such distributions but only a few are needed in practice. These are:

- binomial;
- Poisson;
- geometric; and
- negative binomial.

The binomial occurs as follows: If n independent trials are carried out where the probability of a given event occurring on each trial is constant and equal to p, then the observed number, x, of the given events that occur in the n trials has a binomial distribution. Often the occurrence of a specified event is termed a 'success' and the possible number of 'successes' between 0 and n has an associated probability. For example, if a fair coin is thrown twice, the number of 'heads', x, will be 0, 1 or 2 with associated probabilities of 0.25, 0.5 and 0.25. Here $n = 2$ and $p = 0.5$, x being a binomial variable, so-called because the probabilities of $x = 0, 1, 2, \ldots n$ are given by the successive terms of the binomial expansion of

$$p(x = k) = \binom{n}{k} p^k (1 - p)^{n-k} \tag{2.9}$$

The Poisson is the distribution of the random event. If events occur 'at random' at an underlying mean rate of λ per unit interval, (usually of time or space), then the actual number of events, x, which will occur in any given unit interval has a probability associated with it given by the Poisson model. The probability that there will be exactly x random events in an interval when the mean number is λ is given by

$$P(x) = \frac{e^{-\lambda}\lambda^x}{x!}, \quad x = 0, 1, 2. \ldots \tag{2.10}$$

The geometric distribution is an example of a 'waiting time' distribution and is closely related to the binomial. If, instead of carrying out n binomial trials and observing the number of successes x, trials are continued until the first success occurs, then the number of trials necessary is now the random variable and has a probability associated with it. An obvious extension considers the number of trials necessary to observe exactly k successes; this is known as the negative binomial distribution. The geometric distribution is the special case when $k = 1$.

2.3.7 Continuous probability distributions

The normal distribution is the most well-known of these and it occupies a central place in statistical theory. An important property relates the area under the normal curve to the deviation from the mean in multiples of its standard deviation. Another important distribution in statistical modelling is the negative exponential distribution. This is complementary to the Poisson distribution and describes the lengths of intervals between successive random events occurring at an underlying mean rate of λ per unit interval. The relationship of these two distributions is similar to that between the geometric and the binomial. The negative exponential distribution is used as a basis in reliability modelling and in queuing theory.

2.3.8 Probability density functions

If a sufficiently large number of observations of a continuous variable is made and plotted as a histogram, we can expect the histogram's overall shape will approximate to a smooth curve. In many cases, given some basic modelling assumptions about the population, an equation can be derived for this 'envelope'. Such an equation is called a probability density function (pdf) and is denoted by $f(x)$. The density function or pdf for short, does not directly give event probabilities; it has to be integrated over a range of the variable, x. That is, the probability that x falls in a given range of values is found from the area under the density function between the limiting values, a and b say, of the range. In symbolic terms, this may be stated as:

$$p(x = k) = (1 - p)^{k_p} \tag{2.11}$$

Where $f(x)$ is defined to be zero outside the range a to b. If a and b are the lower and upper limits of the range of x, then the integral (the total area under the curve) is always unity. The density function contains 'normalising factors' to make this happen. Thus the area under any part of the density function will directly give the probability that an observation will fall in that interval.

2.3.9 Cumulative distribution function

The cumulative distribution function (cdf) does return a probability, namely the probability that a randomly sampled value of x will be less than or equal to a specified value.

$$F(x) = \int_a^x f(x)\,dx = \int_{-\infty}^x f(x)\,dx \tag{2.12}$$

2.3.10 Sampling distributions

All of the distributions so far are sampling distributions – they describe what happens to a given variable with indefinite repeated random sampling from a defined population. There are many others, some of which are frequently used. These common ones are the Z (standardised normal), t, F and χ^2 distributions. They are mostly used in hypothesis testing and in confidence interval estimation.

2.3.11 Central limit theorem

One of the reasons why the normal distribution is central in statistical theory is because of this theorem. This states that the distribution of a linear combination of random independent variables will tend to be normally distributed as the number of variables increases, whatever the distributions of the component variables. This holds under fairly general conditions which are usually met in practice. Many sample statistics are linear combinations (or are approximately so) of the sampled variable. The distribution of the sample mean, $\bar{x}$, tends to be normally distributed for sample sizes even as low as 3 or 4 from virtually any population likely to be met in practice. Also, as the number of trials, n, gets large (such that $np > 5$), a binomial variable will tend to be distributed normally. A Poisson variable with a mean greater than about 10 will also be reasonably well approximated by a normal distribution.

2.3.12 Relationships between variables – correlations

It is often useful to evaluate relationships between variables. This can be done graphically using a scatter diagram and/or by the calculation of a regression line. In the latter case, any number of independent variables can in theory be accommodated to assess their influence on the observed response variable.

Regression models are widely used (and abused) and are especially relevant in response surface modelling where, for example, optimum processing conditions are sought between multiple variables. The degree of association between two variables can be summarised by calculating their correlation coefficient. The most common one is the Pearson product-moment version; though there are others based on ranks rather than measured values. There are many pitfalls in the interpretation of correlation coefficients. These traps relate to judging both the statistical and practical significance of the size of an observed coefficient and in assuming that the correlated variables are causally related.

2.4 THE NORMAL DISTRIBUTION

Histograms of observed data are often found to have a particular shape; they are roughly symmetrical with a maximum frequency in the middle and with individual frequencies tailing off on either side of this maximum. This shape is very like a well-known theoretical distribution, namely the Normal or Gaussian distribution, which has some useful mathematical properties. The Normal distribution is shown in Figure 2.1.

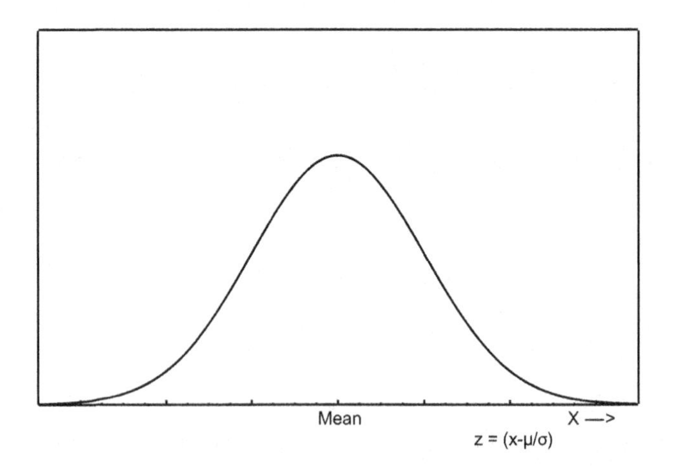

Figure 2.1 The Normal distribution.

2.4.1 Area under the normal curve

If any random variable, called x say, has a Normal distribution with given mean (μ) and standard deviation (σ), then if one random value of x is selected, the chance that it will have a numerical value in any specified interval, can be found by integration. However, the density function of the normal distribution is such that it can not be integrated analytically for arbitrary integration limits. A numerical method of integration is needed. This is done by 'standardising' the observed value x with respect to its distribution's mean and standard deviation and then making reference to a table of areas under the Normal curve. In this way, the numerical only needs to have been done once.

2.4.2 Standardising a normal curve

In symbolic terms, this operation is:

$$\text{Standardised value } Z = (x - \mu)/\sigma \qquad (2.13)$$

Thus, a value of x is expressed as a deviation from the mean, measured in multiples of the standard deviation. A negative Z-value corresponds to a value of x below its mean. Thus, there is a unique correspondence between a value of x and its associated Z-value and either can be found from the other, if μ and σ are known.

2.5 STATISTICS AND PROBABILITIES FOR PROCESS PROBLEMS

2.5.1 Theoretical distributions of failure times

One of the requirements for process systems is to understand process failure, and particularly the time to failure (TTF). This may be the failure of a single component or of a system. Assuming the testing of the asset or system starts at time zero, the failure time of an item is a positive number. We require a continuous theoretical distribution as an appropriate statistical model for the distribution of failure times. Such a distribution can be described either by its probability density function as a function of time $f(t)$ or by its cumulative distribution function $F(t)$ as follows:

EXAMPLE 2.2 MEAN LIFE TIME TO FAILURE

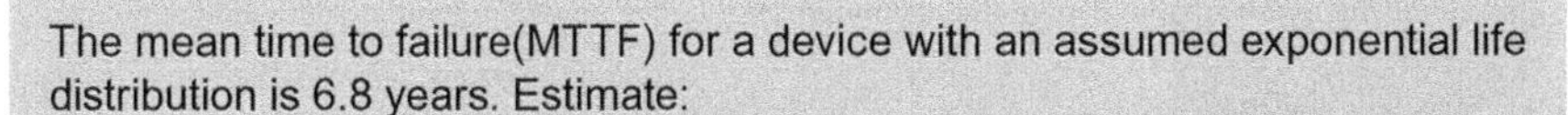

The mean time to failure(MTTF) for a device with an assumed exponential life distribution is 6.8 years. Estimate:

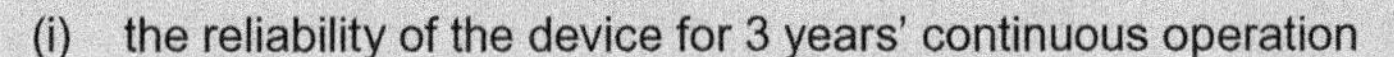

(i) the reliability of the device for 3 years' continuous operation

$$S(t) = \exp(-\lambda t);\ \lambda = 1/\text{MTTF}$$

$$S(t) = \exp[-3 \times (1/6.8)]$$

(ii) the probability that the device will fail before the end of the fith year:

$$F(t) = 1 - \exp(-\lambda.t)$$

$$F(5) = 1 - \exp[-5 \times (1/6.8)]$$

(iii) the probability that the device will fail between the end of the fifth and the end of the seventh year

$$F(7) - F(5) = [\exp(-5 \times 1/6.8) - \exp(-7 \times 1/6.8)]$$

If Δt is a very short time, the probability that an item will fail between times t and $t + \Delta t$ is:

$$f(t) \cdot \Delta t \qquad (2.14)$$

The probability that an item will fail between times 0 and t is:

$$F(t) = \int_0^t f(u)\,du \qquad (2.15)$$

The function $f(t)$ describes a curve, whilst $F(t)$ is the area under that curve. We define $f(t)$ such that the total area under the curve, $F(\infty)$, is equal to one, representing the fact that an item must fail at some time during its asset lifetime. Using calculus,

Credit: istockphoto © eyjafjallajokull

given either f or F, we can construct the other. For mathematical convenience, we introduce two other functions related to $f(t)$ and $F(t)$. The survival function (also called the reliability function):

$$S(t) = 1 - F(t) \tag{2.16}$$

which is the probability that an item will survive until time t, or later, and the conditional failure rate function:

$$Z(t) = \frac{f(t)}{S(t)} \tag{2.17}$$

which is called the hazard function, or the instantaneous failure rate function. This is important because the quantity:

$$Z(t) \cdot \Delta t \tag{2.18}$$

gives us the probability that an item will fail between t and $t + \Delta t$ given that it has survived until time t. Given that $S(t)$ is always less than 1, $Z(t)$ will always be greater than $f(t)$. For example, the proportion of cars that fail between 15 and 16 years is small because few cars last as long as 15 years. But, of the cars that do last 15 years, a substantial proportion will fail in the next year. Thus $Z(t)$ allows us to compute conditional probabilities, hence its name, the conditional failure rate function.

2.5.2 A general link between *Z(t)* and *S(t)*

As mentioned above, $f(t)$ and $F(t)$ are related according to:

$$f(t) = \frac{d}{dt}F(t) \tag{2.19}$$

and because $F(t) = 1 - S(t)$

$$f(t) = -\frac{d}{dt}S(t) \tag{2.20}$$

Thus, inserting this in the definition of $Z(t)$ gives:

$$Z(t) = -\frac{\frac{d}{dt}S(t)}{S(t)} \quad Z(t) = -\frac{d}{dt}(\ln(S(t)) \tag{2.21}$$

$$S(t) = \exp\left(-\int_0^t Z(t)\,dt\right) \tag{2.22}$$

We can now make some assumptions about the conditional failure rate function $Z(t)$ then work out the equivalent $S(t)$, $F(t)$ and $f(t)$.

2.5.3 The simplest *Z(t)*

Suppose a product's age has no affect on the probability of it failing. Then:

$$Z(t) = \lambda \tag{2.23}$$

where λ is a constant value. Using the above relations, this means that

$$S(t) = \exp(-\lambda \cdot t) \tag{2.24}$$

$$F(t) = 1 - \exp(-\lambda \cdot t) \tag{2.25}$$

$$F(t) = \lambda \cdot \exp(-\lambda \cdot t) \tag{2.26}$$

This particular $f(t)$ is called the negative exponential distribution. The figure below shows the exponential probability density functions for $\lambda = 1.5$ and $\lambda = 3$. A component described by the exponential distribution should not be maintained preventively.

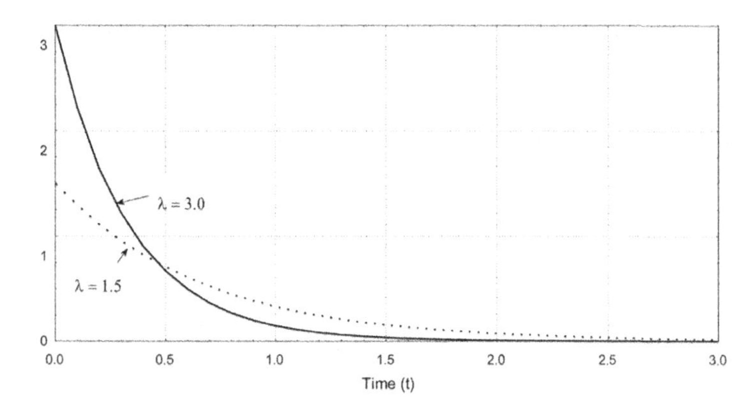

Figure 2.2 The exponential distribution.

The exponential distribution describes a situation where the probability that a component will work over a specific period does not depend on how long the component has worked. It is not a good model for many failure time distributions, since most products fail due to wear-out mechanisms that involve ageing. Such products require more complex forms for $Z(t)$ to produce a useful model. However, the exponential distribution does model the failure of cathode ray tubes and electric fuses adequately.

2.5.4 Estimating the mean lifetime

The mean of the exponential distribution is given by:

$$\int_0^{\infty} t \cdot f(t)dt = \frac{1}{\lambda} \tag{2.28}$$

the reciprocal of the mean rate of occurrence of the underlying random events. The mean lives in the two exponential distributions shown above are 1/1/5 and 1/3, equal to 0.67 and 0.33 respectively. If all the items in a life test sample are run to destruction without replacement, and the exponential model is believed to be useful, then the mean lifetime $1/\lambda$ is estimated from the sample data simply by taking the average of all the sample failure times (see Example 2.2). However, if the test was time-truncated or sample-truncated, or if a replacement scheme is used then the estimate is different.

2.5.5 A different form for *Z*(*t*)

Suppose that $Z(t)$ is taken as:

$$Z(t, \lambda, m) = m \cdot \lambda \cdot t^{m-1} \tag{2.29}$$

where λ and m are constant values.

Proceeding as before, we can work out $S(t)$, $F(t)$ and $f(t)$ for the chosen $Z(t)$:

$$S(t) = \exp(-\lambda t^m) \tag{2.30}$$

$$F(t) = 1 - \exp(-\lambda t^m) \tag{2.31}$$

$$f(t) = m\lambda t^{m-1} \exp(-\lambda t^m) \tag{2.32}$$

This particular density function, $f(t)$, is the Weibull distribution with two parameters. If we substitute $m = 1$, then $f(t) = \lambda\exp(-\lambda t)$ showing that the negative exponential distribution is a special case of the Weibull distribution. The constant, m, is called the shape parameter, while λ is called the scale parameter.

The Weibull model is quite versatile, its density function, $f(t)$, taking on a wide variety of shapes depending on the choices of the values of the parameters m and λ. A yet more general form of the Weibull distribution has a third parameter – one of location. Figure 2.3 shows the range of shapes which the Weibull distribution may take on.

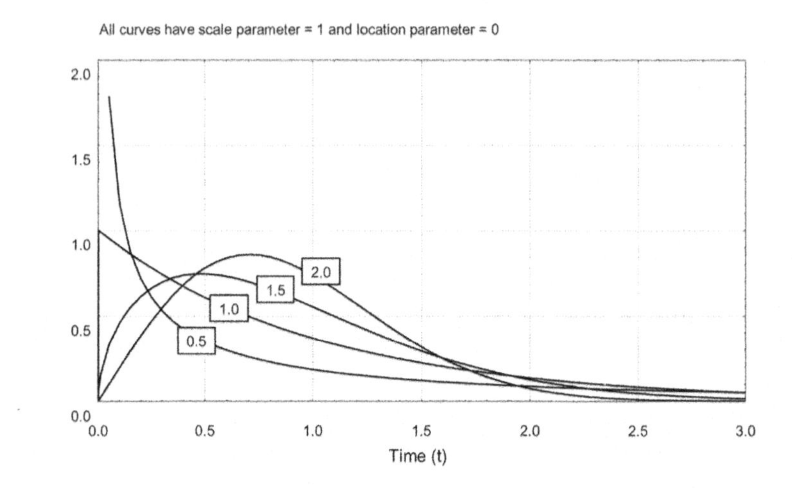

Figure 2.3 Weibull Distribution, shape parameter $m = 0.5$. 1.0. 1.5 and 2.0.

2.5.6 Estimation of mean lifetime

Writing $m = \beta$ and $\lambda = 1/(\alpha\,\beta)$ we can rewrite the Weibull distribution as:

$$f(t) = \frac{\beta}{\alpha^{\beta}} \cdot t^{\beta-1} \cdot \exp\left[-\left(\frac{t}{\alpha}\right)^{\beta}\right] \tag{2.33}$$

This is useful because the mean of the Weibull distribution, defined as:

$$\int_{0}^{\infty} t \cdot f(t)\,dt \tag{2.34}$$

then turns out to be:

$$\text{mean}(\beta,\alpha) \equiv \alpha \cdot \Gamma\left(1 + \frac{1}{\beta}\right) \tag{2.35}$$

where Γ is the so-called gamma function. The mean is not a simple function of m and λ.

2.6 SUMMARY AND SELF-ASSESSMENT QUESTIONS

We have introduced basic statistical and probability concepts in this chapter to a level of sophistication suitable for postgraduate Masters students. We are now ready to consider application of these concepts to unit processes. In doing so, we are principally concerned with design safety and understanding the risks that unit process and system failure may incur – risks to the provision of safe drinking water, to the efficacy of wastewater treatment and the subsequent consequences of failure and, in particular, to public health and the environment. Having understood process risk analysis, we will explore other, less quantifiable, risks and the tools and techniques available for evaluating their significance.

SAQ 2.1 If the probability of failure of an alarm device is p, calculate the probability that at least one out three alarms will fail.

SAQ 2.2 Assume that the probability of a pathogen infecting an exposed individual is p. If n pathogens are ingested and they infect independently of one another, calculate the probability of infection.

SAQ 2.3 Why is preventative maintenance not appropriate for components displaying an exponential life distribution?

2.7 FURTHER READING

There is a range of basic statistical tools on the web. One can be found at: http://www.statstutor.ac.uk/about/

Aven T. (2012). Foundations of risk analysis: A knowledge and decision oriented perspective, 2nd edn, John Wiley & Sons, Chichester, UK.

Berthouex, P. M. and Brown, L. C. (2002). Statistics for environmental engineers, 2nd edn, Lewis Publishers, MI, US.

Christodoulou, S. E. (2011). Water network assessment and reliability analysis by use of survival analysis. *Water Resources Management*, **25**, 1229–1238.

Martínez-Rodríguez, J. B., Montalvo, I., Izquierdo, J. and Pérez-García, R. Reliability and tolerance comparison in water supply networks (2011) *Water Resources Management*, **25**, 1437–1448.

Paustenbach (ed.) D. J (2002). Human and ecological risk assessment: Theory and practice, Wiley-Interscience, New York, US.

Vose D. (2008). Risk analysis: a quantitative guide, 3rd edition, John Wiley & Sons, Chichester.

Unit 3

Process risk and reliability analysis

SOME KEY TERMS FOR WATER AND WASTEWATER TREATMENT ASSETS

reliability – the probability that an asset is able to perform a required function under stated conditions for a stated period of time or for a stated demand.
availability – the probability that an asset is in a state available to perform a required function.
maintainability – the property of design that determines the effectiveness with which an asset can be restored back to a state in which it can perform its required function.

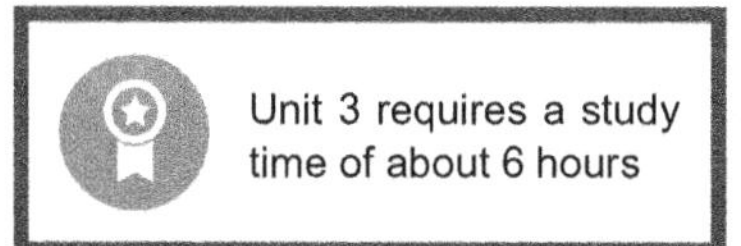

3.1 CONTEXT

We now consider the practical application of risk assessment and management in the water and wastewater utility sector. Units 3, 4 and then 5 and 6 are concerned, progressively with (i) risks *within* the process plant boundary, within the 'factory fence' if you will; (ii) risks *outside* of the process plant boundary in the wider environment; and then (iii) business risks and their regulation *within* the organisation, or the 'body corporate'. This progression follows the growing application of risk thinking within the process sector and also the development of powers that regulators have progressively assumed through legislation to oversee water sector risks. A short-hand description of the material in these units is engineering risk and reliability (unit 3), environmental risk assessment (unit 4) and regulatory and corporate risk management (units 5 and 6). Readers will recognise this moves us from analysing the performance of closed engineered systems, through to their interface with open natural systems and, ultimately, to the preserve of human and organisational systems that interface with those of other stakeholder bodies. As we progress through this series, we recognise a loss of direct operator control and a rise in what commentators are increasingly calling 'extrinsic risk'.

Having introduced risk analysis and the basic statistical and probability tools, we can now explore their use, initially in the context of evaluating the risks associated with unit treatment processes. Following this unit, our discussion will broaden to consider the wider aspects of risk assessment and management that are critical to the management of water and wastewater utilities, but for now we are concerned with unit processes themselves.

EXAMPLE 3.1 ESTIMATING ASSET AVAILABILITY

For utility operations, the time taken to restore operations following a failure or operating problem can be significant. Under steady state conditions, the asset availability (A) is given by:

$$\frac{\text{mean time between failure (MTBF)}}{\text{MTBF + mean time to restore operation}}$$

that is, the ratio of the operating time to the total time.

Before applying some of the theory from Unit 2, it is helpful to consider the context of application, in order to ground their use in reality. One way of viewing treatment processes is as organisational assets that we use in order to protect public health and the environment. By operating these assets, (water treatment plant, reservoirs, wastewater treatment processes, distribution networks), often in a highly optimised fashion, we put the assets at risk in order to deliver safe drinking water or acceptable wastewater discharges. The assets must be designed so that they are inherently safe to operate as large pieces of physical, chemical or biochemical plant and so that if failures were to occur, there is sufficient spare capacity or system 'redundancy' to avert a major incident. This is the concept, in practice of maintaining a stand-by or reserve pump, for example.

We are concerned then with the potential probabilities and consequences of failure in design as well as in the consequences of unit process and system failure.

3.2 APPLYING PROCESS RISK ANALYSIS

For now, we can consider risk assessments as providing an estimate of risk, based on a combination of the probability and consequences of specified events. Risk management is then the process by which the results from risk assessments are used to make decisions about reducing or mitigating the potential consequences.

Having understood and articulated the problem under study, the starting point of any risk assessment is the identification of hazards. Hazard identification involves the rigorous consideration of all situations in which the potential for harm may exist. It can include a disciplined analysis of the combination or sequences of events that could transform this potential into an incident.

Consider the likelihood and consequences of a unit process failure – say the process failure of a biosolids pyrolysis plant. Normally, for unit processes, consideration is given to the following three aspects of hazard identification:

(i) determining whether a given operation or activity has the potential to give rise to a hazardous event;
(ii) determining the range of hazardous events that the operation or activity could present; and
(iii) identifying the routes by which each of these hazardous events could be realised, *i.e.* identifying the potential incident scenarios.

Effective hazard identification and subsequent risk assessment depends on intimate practical knowledge and experience of the process being identified, and uses a systematic method to ensure that this knowledge and experience is applied effectively. To aid application, a number of hazard identification techniques have been developed. Creativity is important here so one does not become locked-in to a particular mindset, or a restricted number or type of hazards. Selection of the appropriate technique enables the analyst to focus on the particular issues of concern and allows the analysis to be undertaken to an appropriate level of detail. Two main types of hazard identification technique are discussed below:

(i) **comparative** methods – these draw on knowledge gained from experience and include methods such as checklists and hazard indices.

THE MULTIBARRIER APPROACH

The accepted approach to water treatment is the multibarrier approach. This usually refers to an integrated system of procedures, processes and tools that collectively prevent or reduce the contamination of drinking water from source to tap in order to reduce risks to public health. Water treatment processes are one part of the approach, together with source water protection, operation of a secure water distribution network and appropriate customer use of water.

The central premise of the approach is a purposeful reliance of more than one barrier as a means of protecting water customers from harm. Multiple barriers afford logarithmically greater levels of protection by separating the source of the hazard from the receptor. For water treatment process, active maintenance and asset renewal plays a central role.

(ii) **fundamental** methods – these are structured ways for stimulating and encouraging individuals or groups to apply creative foresight, based on their knowledge and experience, to the tasks of identifying hazards by using a series of 'WHAT IF' questions.

Having discussed these, we will progress with some quantitative problems and examples.

3.2.1 Comparative methods

Checklists represent an important means for readily identifying hazards. They may be derived from experience alone, (including operational codes of practice or standards), or for a particular type of generic plant from previous applications of the fundamental techniques. For example, a water utility may be able to apply the results of a previous HAZOP study, (see below), within a checklist for the hazard identification on another treatment works.

Credit: istockphoto © pupunkkop

A variety of checklists have been developed for application in the process industries covering many aspects of design, commissioning, operation and decommissioning, as well as safety related features. Any checklist is necessarily general and will not be exactly applicable to a particular plant, process or issue. In spite of this, checklists can be extremely useful in identifying quickly a large range of hazardous events. The main disadvantage with them is that, to be comprehensive, they may be cumbersome. They may also inadvertently discourage creative thought by leading the user to assume that all aspects that ought to be questioned have been covered in the checklist, and so some serious hazards may be missed. What they do not provide is a fundamental examination of the hazard. This is an important drawback in cases of novel design or in unusual situations where there may be no previous experience to draw on, and here, a more fundamental approach is required.

3.2.2 Hazard indices

Hazard indices are designed to give a semi-quantitative indication of the relative potential for hazardous incidents associated with a given plant or process. One such hazard index is the Dow Index, designed originally to guide the selection of fire protection methods. Three empirical hazard factors are determined:

(i) materials hazard factor;
(ii) general process hazard factor; and
(iii) special process hazard factor.

TYPICAL CONSTITUTION OF A MULTI-DISCIPLINARY HAZOP TEAM

- HAZOP chairman
- HAZOP secretary
- design engineer
- operations manager
- process operator
- maintenance manager
- process scientist
- risk analyst
- health and safety manager

In the Dow Index, the materials hazard factor, ranging between 1 and 40, is a reflection of the properties of the materials being handled. It gives an indication of their energy potential, including mixtures, whenever a hazardous material is present in sufficient quantity in the processing plant to present a hazard. The materials hazard factor is then modified by the other two factors that act as weighting factors. The resulting numerical value provides a 'fire and explosion index' for the process under study. This index enables the hazard to be estimated by comparison with the value of the index for process units of known hazard potential. The Dow Index categorises hazards overall as light, moderate, heavy or severe.

The use of a hazard index in the early design stages of a project may reveal a hazard potential that could then be reduced before the design is advanced. However, the use of such indices is more suited to the design stage of plant rather than for hazard identification on operating processes. Process hazards relating to items such as novel processes, poor plant management, inadequate design, human error, *etc.* cannot be accounted for adequately in this type of index. As a result of these limitations, hazard indices are not normally used for hazard identification on existing plant. Effective use is limited to the early design stage for new plant, and even then only as a rapid screening tool.

3.2.3 Fundamental methods – qualitative

The hazard and operability study, or HAZOP, is an established technique for systematically considering all possible deviations from the design intent of process plant, (and their consequences), by the application of a series of 'guide words' recognisable to process engineers. A HAZOP is used to examine new plant or system designs as well as existing plant to identify potential hazards and problems that may prevent efficient or safe operation. For the water and wastewater utility sector, HAZOPs are usually the preserve of design, chemical and process safety engineers, who design, build and operate process plant. Modern process engineering software packages are available that facilitate the HAZOP process on desktops. Integrated software that links HAZOP to process flowsheets, SCADA (supervisory control and data acquisition) systems and 2-dimensional/3dimensional computer-aided design (CAD) packages are now commonplace in water and wastewater treatment control rooms.

EXAMPLE 3.2 DESIGNING AND MAINTAINING SYSTEMS RELIABILITY

Compagnie Generale des Eaux (CGE) applied an integrated approach to evaluating plant reliability to a post-chlorination system in their Neuilly-sur-Marne plant. CGE defined the system in reliability block diagrams and modelled the risks based on hazards identified from a HAZOP study. The risk consequences are quantified in a FMECA study. The computation of system availability is the probability of the system being operational.

The availability study based on HAZOP, FMECA supported the quantification of acceptable risk in designing a reliable and available system for operation. Reliability analysis during conceptual design determines acceptable risk and cost early in the life cycle of assets but also determines maintenance procedures to maintain design quality.

The HAZOP technique relies on the imagination of a multi-disciplinary team to consider all of the ways in which a plant might deviate from its normal operating conditions. The team comprises individuals who, together, will have a comprehensive knowledge of the process being studied, including the intended behaviour of the process and the operating procedures. The technique is applied systematically, following a predetermined plan, a process that requires competent leadership from a HAZAOP-trained manager. The team focuses on each sub-system in turn and considers a list of parameters and guide words for the sub-system to identify all ways in which the plant could deviate from its design intent. For each deviation identified, the team assesses its consequences and, if appropriate, recommends the corrective action that should be undertaken.

The HAZOP technique was developed by the former chemical giant Imperial Chemical Industries (ICI) to assess the potential hazards presented by large, continuous, single stream chemical plants to their operators and to the public (as with all risk assessment studies, maximum benefit is obtained when the study is performed during the design stages, when it is easier to make changes without incurring too much additional expenditure). This HAZOP technique is now widely used throughout the utility sector. In the early stages of a project, a preliminary hazard analysis acts as a high level examination with the objectives of identifying any:

- potential hazards or operability problems associated with a system, process or procedure;
- relevant codes or standards applicable; and
- areas of uncertainty or additional safeguards that may be required.

The study requires the sub-division of the overall system or process into a series of sub-systems or nodes. These are then examined in turn by applying a series

PROCESS SAFETY ENGINEERING ROLES IN THE UTILITY SECTOR

The utility sector has developed the role of process safety engineer. These professionals are required to undertake the full range of qualitative and quantitative process risk analysis tools. Employed by utilities directly, or supplied by professional engineering consultants, the process safety engineer will typically apply HAZOP, FTA and FMECA analyses to process plant design and have a deep understanding of the regulatory environment, hazardous chemicals, and processes, working practices, and ergonomics/human factors in plant design and operation.

of generic keywords that are applicable to the operation of the sub-system, (such as fire, smoke, explosion, contamination, criticality, loss of ventilation, loss of services). The study team considers whether there are any situations or initiating events that may lead to these hazardous key events. Once the full list of relevant keywords has been considered, the team considers the next node, repeating the overall process until all nodes have been examined. The preliminary hazard analysis is best initiated as soon as the necessary information is available, e.g. operational process flow sheets, location and citing of plant, and building layouts. Along with all risk assessments, once the detail becomes available, it is appropriate to iterate the analysis and undertake a detailed HAZOP study.

Only when major hazards and areas of high risk for a system are identified can action be taken to reduce the risks of failure. The use of HAZOP studies allows the impact of operational or design changes on the plant to be assessed before decisions are made. Thus, when changes are made, management can be sure that those changes are optimised within the given constraints. Optimisation can be carried out with respect to a variety of criteria, for example:

- where performance can be improved within the system, (*i.e.* the removal of pinch-points in the design);
- whether to install more redundancy, use more expensive, (but more reliable), equipment, or change operating procedures; and
- where to locate resources such as repair staff and spares and how to use them to best effect.

Failure modes and effects analysis (FMEA) is another qualitative technique. FMEA evaluates system design by examining component failure modes and operating regimes, with the objective of determining the effects of malfunctions and failures. By tracing the effects of individual component failures up to the system level, the criticality of particular failures, in terms of their consequences, can be assessed. This allows risk-based corrective action to be prioritised and taken to improve the design by determining ways to eliminate or reduce the probabilities of occurrence of critical failure modes. The approach can also be used for examining existing plant, e.g. to identify the need for the introduction of new safety procedures.

For the most basic examination, a failure mode analysis (FMA) is all that is required. Generally, however, the subsequent effect of each failure mode on the process system is considered – a failure mode and effects analysis (FMEA). The approach can then be extended further to include an assessment of the follow-on consequences of each effect, if it is realised. This is achieved by ranking the consequences of failure, with the resulting approach being termed a failure mode, effects and criticality analysis (FMECA).

The procedure is highly flexible and applied in water and wastewater treatment, both in assessing individual items of plant and ranking the risks associated with each process element within the whole treatment system. The overall approach has similarities with HAZOP, with keywords being used to aid the identification of failure modes.

A further qualitative tool is the hazard analysis and critical control point methodology (HACCP; Havelarr, 1994; Hellier, 2000). HACCP is important in that it has informed the preparation of water safety plans (WSP) in the revised drinking water guidelines of the World Health Organisation (WHO, 2004, 2011). HACCP has its roots as a food process control system for the management of microbiological risks to consumer food products. The methodology is used extensively within the food sector, (including storage and distribution), to target and reduce harmful biological, chemical or physical contaminants. Its application can also verify that control systems are working as intended to minimise or eliminate these contaminants. HACCP applied to the production of drinking water adopts the basic 7 step procedure:

(1) Conduct a hazard analysis. This involves the preparation of a list of steps in the process where significant hazards may occur and a description of any preventive measures.
(2) Identify the Critical Control Points (CCPs) in the process.

(3) Establish critical limits for preventive measures associated with each identified CCP.
(4) Establish CCP monitoring requirements, together with procedures for using the results of monitoring to adjust the process and maintain control.
(5) Establish corrective actions to be taken when monitoring indicates that there is a deviation from an established critical limit.
(6) Implement effective record-keeping procedures that document the HACCP system.
(7) Establish procedures to verify that the HACCP system is working correctly.

Key components of a drinking **water safety plan** (after WHO, 2004, 2011):

- a **system assessment** to determine whether the drinking-water supply chain (catchment to tap) can deliver water of a quality that meets health-based targets. This includes an assessment of design criteria for new systems.
- identifying **control measures** in a drinking-water system that will collectively control identified risks and ensure that health-based targets are met.
- for each control measure identified, an appropriate means of **operational monitoring** should be defined that will ensure that any deviation from required performance is rapidly detected in a timely manner.
- **management** plans describing actions to be taken during normal operation or incident conditions and documenting the system assessment (including upgrade and improvement), monitoring and communication plans and supporting programmes.

These seven principles were established by the National Advisory Committee for the Microbiological Criteria for Foods (NACMCF) in 1992. Since then, the US Food and Drug Administration (FDA) and the US Department of Agriculture, Food Safety and Inspection Service (FSIS) have put together regulations requiring the use of HACCP in the seafood, red meat and poultry industries.

The most applicable of the HACCP processes to water treatment is likely to be: **receive – hold – prepare – hold – serve**; where the 'hold' stage would be applicable to raw or treated water storage. By sub-dividing 'prepare' into the stages of water treatment, (abstraction – coagulation – clarification – filtration – disinfection – distribution), the hazard analysis part of the HACCP process begins to resemble the HAZOP or FMEA techniques above. However, HACCP also has the additional statistical process control and quality assurance procedures to ensure that the process operates consistently within the required parameters.

The European water industry, in collaboration with the American Water Works Association and the Australian water industry, reviewed the potential for applying HACCP across the whole of water supply, from catchment to tap. It is not appropriate, or practical, to simply transfer HACCP wholesale without adaptation, but the key principles underlying the HACCP approach are now adopted within water safety plans. These are concerned with (i) applying preventative risk management from catchment to tap, rather than with unit processes in isolation; and (ii) identifying key risk drives and therefore critical control points for risk management for the entire water supply chain.

Rapid ranking is a simple technique used to gain an appreciation of which hazards are of greatest concern and require further investigation. The technique is undertaken following the application of some form of hazard identification technique such as HAZOP or FMEA. It has been used where a large number of hazards or hazardous sites have been identified and there is insufficient resource to deal with all of them. In these situations it is important that the hazards presenting the greatest risks are identified quickly and easily. Rapid ranking provides a suitable method to achieve this.

A team of experts (usually the same team who undertook the qualitative HAZOP or FMEA study) rank the hazards in terms of their estimated probabilities of occurrence and potential consequence, according to a series of attributes or characteristics. The number of categories depends on the requirements of the exercise but five categories are normally used. An example of how they might be defined is shown below:

Table 3.1 Ranking table, with example likelihoods and consequences defined.

Likelihood Factor		Consequence Factor	
1	**Very unlikely** Less than one per 100 years	1	**No impact** Increased operator effort only)
2	**Unlikely** More than one per 100 years	2	**Limited impact** Minor deterioration in output quality leading to internal incident report
3	**Moderate** More than one per 10 years	3	**Moderate** Customer awareness, increased pressure group action
4	**Frequent** More than one per year	4	**Severe** Regulatory exceedance, adverse publicity
5	**Very frequent** More than one per month	5	**Catastrophic** Life threatening, major environmental damage

Using such a matrix, a risk rank can be allocated to each risk. Risks can then be ordered according to their associated matrix value. If required, the figures within the matrix can be weighted to take account of group or company attitudes to risk, with risk averse companies giving higher weightings to the high consequence events, but these levels of sophistication have their drawbacks.

Credit: istockphoto © jazz42

Table 3.2 Risk matrix, with numerical surrogates for likelihood and consequence.

		Likelihood				
		1	**2**	**3**	**4**	**5**
Consequence	5	5	10	15	20	25
	4	4	8	12	16	20
	3	3	6	9	12	15
	2	2	4	6	12	10
	1	1	2	3	4	5

The shaded areas of Table 3.2 have their similarities with the ALARP figure in Unit 1 (Figure 1.2) and many organisations have used these matrices as 'heat maps' to position high (dark shading), medium (grey) and low risks (unshaded) on. Whilst this may offer a convenient communication tool for a range of risks, there is caution here:

- The nominal representation of likelihood and consequence (1, 2, 3 ...) rarely represents differences in actual magnitude between individuals risks and so the numbers can only ever be used as surrogates for material differences in likelihood and consequence.
- It follows that the manipulation of risk scores rarely has meaning, so risk scores should not be added, transformed, or used as the basis for derived qualities (insurance premiums).
- Risk matrices are best kept simple and qualitative, ideally used alone for distinguishes between groups of risk that require immediate attention (high), those requiring further risk assessment (grey) and those deemed acceptable albeit requiring onward monitoring (unshaded).

Fault logic diagrams provide a means of representing failure logic schematically. The main techniques adopting this approach are fault trees and event trees. Fault trees (Figure 3.1) have a vertically-aligned structure and, at the top of them, a system or process failure (the so-called 'top event'), with the contributing, and ultimate root causes set below. Event trees usually adopt a horizontally-aligned structure and adopt a failure as the initiating event, on the left of the page, from which subsequent effects, consequences or outcomes are then manifest. For application within the water and wastewater utility sector, fault trees are of value for assessing the root causes of process failure; and event trees for assessing the implications and onward consequences of failure.

Fault trees encourage the analyst to speculate how a particular hazardous situation could arise. Event trees then focus on what may ensue from such a situation, and thereby identify the outcomes of specific undesired events. However, their main purpose is to provide a structure for the failure logic of events that have been identified by some other method. By providing the logical structure, these diagrams can form the basis of a quantitative assessment where, that is, sufficient data of suitable quality for quantification are available.

In practice, the opportunities for taking fault logic diagrams through to full quantification are relatively limited within the water and wastewater sector. This is due to difficulties in modelling the time dependent benefits of storage facilities appropriately within the constraints of fault and event tree logic. However, where time dependent features such as storage are not an issue, logic diagrams can be a useful method for analysing a system. The basic approach adopted when constructing a fault tree is as follows:

(i) Identify an undesired event, (the top event). This may be system failure or a particular fault condition or an accident which is to be avoided e.g. a fire or explosion.
(ii) By a detailed analysis of the system, identify the combination of events, conditions etc. which bring about the top event. These intermediate events will usually be linked either through logical AND or through Logical OR gates, (described below).
(iii) Continue to work backwards, (deduce) sequentially, identifying all intermediate causes of the events at the higher level.
(iv) The tree stops when no further analysis is possible e.g. a basic event or when further analysis is not practical.

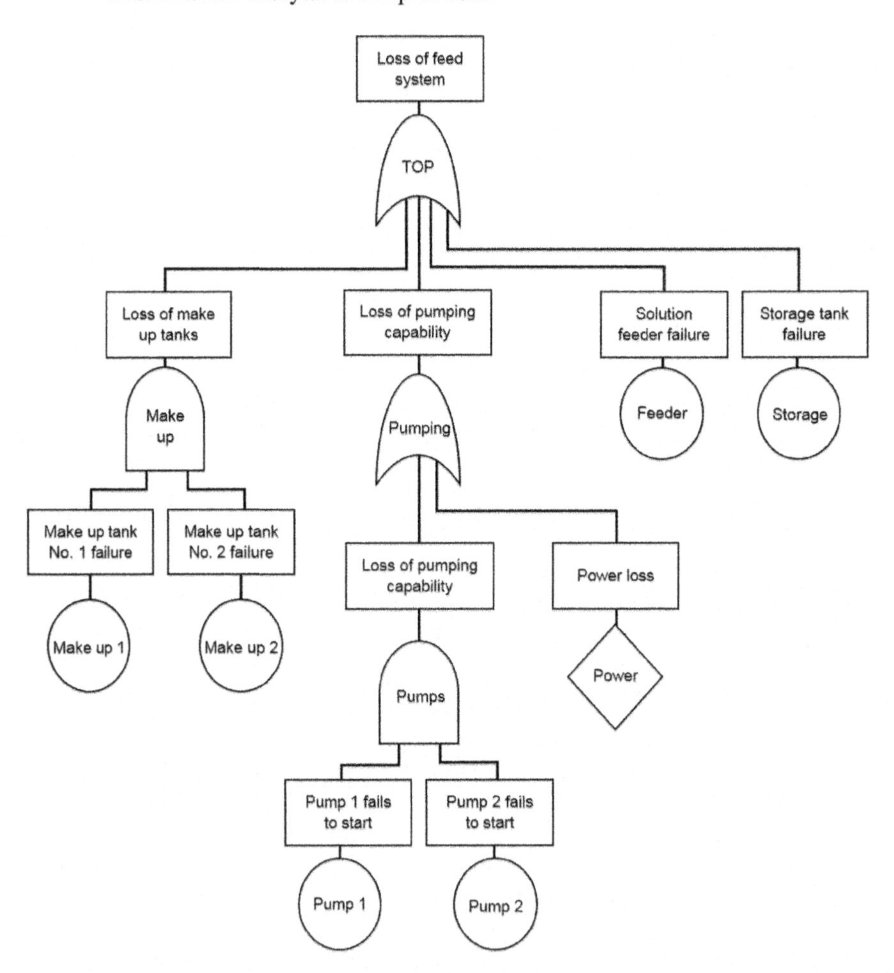

Figure 3.1 Fault tree for a chemical feed system.

Event trees begin with the initiating event and 'map out' the consequences that could arise following a linear sequence of subsequent conditions, each with a probability assigned to the contingency (likelihood) of that condition. The value of event tree analysis (ETA) is usually in describing the set of events as a whole, rather than in quantifying the relative probabilities, but when data is available and both the probability and consequences of events can be evidenced with confidence, then event trees can be used to illustrate key contributors to overall risk and thereby priorities for risk management. Application of quantitative ETA is usually restricted to linear process systems arranged in series where reliable data exists about unit process performance. Useful illustrative examples exist for the analysis of prions, the pathogens that causes transmissible spongiform encephalopathies (TSEs), through the unit processes of wastewater treatment works.

With the increasing interconnectedness of utility systems and infrastructures, risk techniques assuming a simple arrangement of unit processes in linear series will become limited in their usefulness for the future. Complex systems with difficult-to-quantify probabilities, arranged in networks that impact on one another in unpredictable ways, require a new suite of tools that can characterise the complexity of these interactions and interrogate their robustness or vulnerability accordingly.

Table 3.3 Qualitative event tree for a duty/standby sampling system.

Initiating event	Question 1	Question 2	Question 3	Outcome
Failure of duty sample pump	Automatic changeover to standby pump?	Is the standby pump in automatic start-up mode?	Standby pump operates successfully on demand?	Sampling continues successfully
	Yes	Yes	Yes	Yes
			No	No
		No	Yes	No
			No	No
	No	Yes	Yes	No
			No	No
		No	Yes	No
			No	No

DRAWING AN EVENT TREE

A systematic methodology for drawing an event tree involves taking the first component affected by the initiating event, drawing a path for all possible outcomes (consequences), then taking a second component and drawing new branches that represent all possible outcomes. There are 2^n possible outcomes for n affected components.

3.2.4 Fundamental methods – quantitative

The quantitative principles introduced already all allow analytical representations of the system or process under consideration. However, in some instances, the intrinsic limitations of these techniques, or the complex mathematics that would result, may mean that it is not possible to build a representative model. This is particularly so when considering operational issues. Examples of such intrinsic limitations include the following:

- the need to assume constant rates of component failure or repair, i.e. it is not possible to model the component ageing process;

EXAMPLE 3.3 IMPACT OF AVAILABILITY ON CORROSION INHIBITION

Most infrastructures suffer corrosion problems. Inhibitors such as anticorrosion paints, cathodic protection, biocides and inhibitors slow down the rate of corrosion. But their effectiveness (E) is a function of their reliability and availability (A). The corrosion rate, r, is a function of the non-inhibited corrosion rate (r_o) modified accordingly and is given by:

$$r = r_o\,(1 - E\,A_I)$$

If the availability, A_I, of the inhibition system (see example 3.2) is 0.8 and the effectiveness 90%, then a non-inhibited corrosion rate (r_o) of 0.5 mm/yr becomes

$$r = r_o\,(1 - E\,A_I)$$

$$= 0.5[1 - (0.9 \times 0.8) = 0.14 \text{ mm/yr}$$

- an inability to fully model detailed aspects of equipment repair such as spares holding, delays before repair can start, or repair priorities where there are repair resource limitations; and
- an inability to model time dependent features within a system such as lag times or the effect of storage within a system.

Where analytical models are not appropriate, one method that can be used to bypass the complex mathematics and allow more complicated operational and failure or repair patterns to be modelled is Monte Carlo simulation, which has come to the fore with the availability of user-friendly desktop packages.

The success of a Monte Carlo simulation is dependent on being able to generate, mathematically, large numbers of random events that occur within a system. These mathematically-derived events, such as component failure and repair times, can then be manipulated to predict a wide range of system failure information, such as:

- mean Time Between Failures (MTBF);
- mean Time To Repair (MTTR);
- total Uptime during the Study period;
- total Downtime during the Study period; and
- volume of water demanded but not supplied, or sewage dispatched but not received, during the Study period.

Monte Carlo simulation has been used widely for assessing the risks of discharge consent failure to the pollution of water courses and has been useful in some applications within the assessment of water and sewage treatment systems. However, like all quantitative techniques, its usefulness is limited by the paucity of readily available historical data on which such techniques must ultimately depend. Most water utilities have collected vast amounts of data over the years. However, the data have not generally been stored in a form such that they can be easily used within a risk assessment. With implementation of the water safety plan approach, this is beginning to change.

3.3 SYSTEMS RELIABILITY ANALYSIS

It is useful to extend our discussion of quantitative methods and consider systems reliability analysis – principally to understand how linked unit processes behave rather than necessarily for estimating precise systems reliability data. This is helpful because the discussion and examples provide important insights into key aspects of asset management.

An important part of any reliability engineer's work is to predict the reliability of a system from its component parts. Given that the reliability of individual components is available, we are principally concerned with how reliable the system as a whole is. There are a number of well-established techniques available to achieve this task, the best known of which are:

- reliability block diagrams and networks;
- fault tree analysis; and
- event tree analysis.

We have discussed the latter two above. Here we introduce network models and reliability block diagram (RBD) for estimating whole system reliability.

Credit: istockphoto © LiuNian

3.3.1 Reliability block diagrams

In practice it is often found that a system of components, or an activity, can be modelled as a 'network' in which blocks representing components of the system are connected in series or in parallel, rather like electrical resistors. For example, consider a valve as a system. A valve has three major components; the valve mechanism that isolates the pipe system, the actuator that causes the valve to open or close and the control system providing the signal to initiate opening or closing of the valve. Mechanical engineers know that for the valve to operate, each of the three sub components must operate simultaneously. The reliability of a valve system can be represented by three blocks arranged in series.

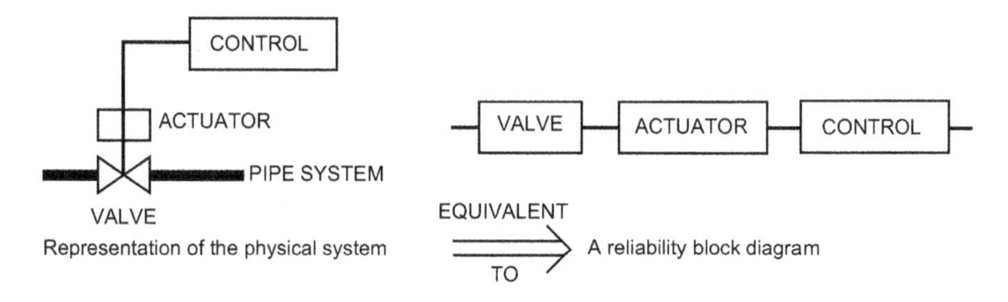

Figure 3.2 Reliability block diagram for valve system.

The reliability block diagram can be thought of as analogous to an electrical circuit, in which each block acts as a kind of conductor. The failure of any one component represented by a block causes the conductor to become 'open-circuit'. Success occurs when there is uninterrupted flow through the network. Conversely, failure occurs when flow through the network is interrupted. The failure of any one of the components will interrupt the flow through the network and cause the system to fail. The relationship between the physical system being modelled and the reliability block diagram is not always intuitive. A physical system, having a well defined topography, may have a completely different reliability block diagram. The RBD arrangement depends on the failure mode under consideration.

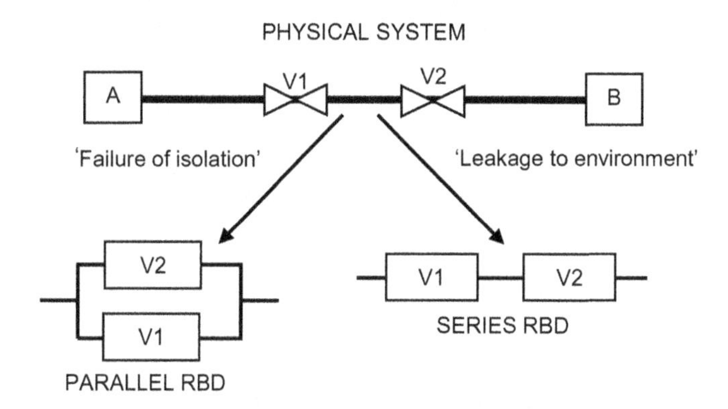

Figure 3.3 Arrangement of a block diagram depends on the failure mode.

Consider the arrangement of components in Figure 3.3, for example. In physical space, the two valves (V1 and V2) are in series on the pipe. The design engineer's intent is to provide a backup such that in the event that one of the two valves fails to close, there is a second that can be closed to shut off the flow. Thus both component valves would have to fail for the two-valve system to fail. The corresponding reliability block diagram (RBD) for a 'failure of isolation' is therefore a parallel arrangement even though the valves are physically placed in series. Notice, however, that if the failure mode 'leakage to the environment' is modelled, then because either valve can leak to the environment, then the RBD for leakage to environment is a series configuration. Next consider a three component system arranged in series as shown in Figure 3.4 below

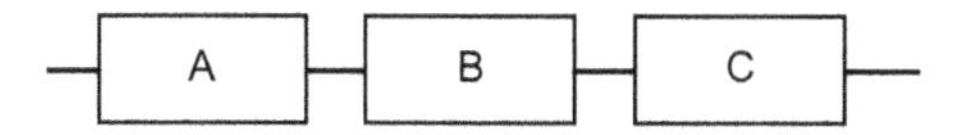

Figure 3.4 A three component system in series.

THE MULTI BARRIER SYSTEM

Here we see the value of the multi barrier system for water treatment, which relies on a series of well-maintained 'interventions', or barriers, to prevent pathogens or chemicals at undesirable concentrations reaching the customer's tap. The more barriers, the greater the overall system performance. However, consider the drastic implications of a single component failing to provide its required level of integrity. In a 5 component system of unit processes arranged in series, even if a single component has 10% reliability, the overall reliability plummets to 8%. There are obvious implications here for the management of preventative maintenance regimes, retaining the integrity of critical systems and proactive asset refurbishment and replacement.

In this case:

- the system is non-redundant;
- A, B and C must work for system success; and
- only one element need fail for the system to fail.

If R_A R_B and R_C represent the reliability or probability of successful operation of components A, B and C and Q_A, Q_B and Q_C represents the probability of failure of A, B and C; then the success of the system (S) can be represented by Boolean logic as:

$$S = A \cap B \cap C \quad (3.1)$$

and the reliability or probability of success of the systems as:

$$R_s = R_A \cdot R_B \cdot R_C \quad (3.2)$$

For n components in series this can be generalised to

$$R_s = R_1 \cdot R_2 \cdot R_3 \cdot R_4 \ldots\ldots Rn. \text{ [product rule]} \quad (3.3)$$

There are two important properties of a series system:

- the greater the number of components, the lower the system reliability;
- the least reliable component (the weakest link), determines the overall reliability of the system.

Consider an example. Take a 10 component series system each component with a reliability of 0.95, the overall reliability of the system (R_s), using the product rule:

$$R_s = (0.95)^{10} = 0.599 \quad (3.4)$$

Compare this figure with that for a system containing only 5 components, each with the same reliability. In this case the reliability calculates as:

$$R_s = (0.95)^5 = 0.774 \quad (3.5)$$

This demonstrates that the greater the number of components, the less the reliability of a system when the system is arranged in series. This is the origin of the engineers' rule of thumb: 'the greater the number of components, the greater the complexity, the more things that can go wrong and the lower the reliability'.

Now let's consider a 5 component series system where four of the five components have a reliability of 0.95 but one being fairly unreliable has a reliability of 0.4. The reliability of the system is calculated as:

$$R_s = (0.4) \cdot (0.95)^4 = 0.326 \quad (3.6)$$

Compared with the result obtained when all 5 components had a reliability of 0.95, giving a system reliability of 0.77, the reliability has dropped significantly and is just lower than the reliability of the least reliable component. This demonstrates the reliability of a series system is dominated by the least reliable component; the weakest link in the chain. If a single component in this system expressed a 10% reliability (0.10), the system reliability falls dramatically to 0.081.

What about parallel networks? Consider a three component system with a parallel arrangement. This can be represented in the diagrammatic form below:

When components or subsystems are arranged in parallel:

- the system is fully redundant;
- providing <u>any one</u> of A or B or C, or combinations thereof, are maintained in a working state, then the system will operate successfully;
- in this case, <u>all</u> components must fail for the system to fail.

If the reliability, or the probability of a successful operation of the individual components A, B and C are written as R_A, R_B and R_C and the probabilities of failure by Q_A Q_s and Q_C respectively, and if the system failure is represented by F, then the failure of the system can be represented in Boolean logic as:

$$F = A \cap B \cap C \quad (3.7)$$

and the probability of failure of the system is given by:

$$Q_S = Q_A \cdot Q_B \cdot Q_C \tag{3.8}$$

$$R_S = 1 - [(1 - R_A) \cdot (1 - R_B) \cdot (1 - R_C)] \tag{3.9}$$

Consider a 3 component system with A, B and C arranged in parallel (Figure 3.5), each with a reliability of 0.7. What is the overall reliability of the system?

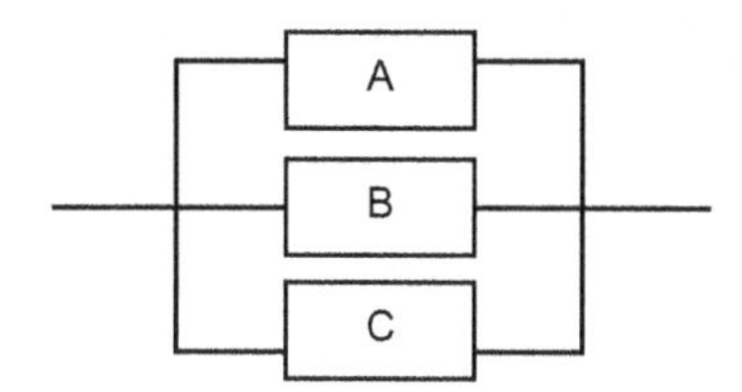

Figure 3.5 A parallel network.

$$Q_S = Q_A \cdot Q_B \cdot Q_C = (1 - 0.7)^3 = 0.027 \tag{3.10}$$

$$R_S = 1 - Q_S = 1 - 0.027 = 0.973 \tag{3.11}$$

Note, the system reliability increases from 0.7, when the system has only one component, to 0.97 when 3 identical components are arranged in parallel. This is the mathematical basis of providing system redundancy (e.g. back-up, or stand-by pumps or assemblies in the event of the critical failure of the duty pump; developed further in section 3.3.5 below).

3.3.2 System redundancy

Suppose it is necessary to measure a particular concentration of a species using a sensor. If the sensor is unreliable, say failures occur 1-2 times per year, (*i.e.* $P = 10^{-1}$ over one year), then the sensor system can be made to have a much lower probability of failure over one year by duplicating or triplicating the sensors. If the system can shut down as a result of a signal from any one of three sensors, i.e. the sensor system fails if all three sensors fail, then the overall probability of failure is:

$$P\ (\text{system}) = p \times p \times p = p^3 = (10^{-1})^3 = 10^{-3} \tag{3.12}$$

The system logic is as follows. Define the events as follows:

E1: no signal from sensor 1 when danger is present
E2: no signal from sensor 2 when danger is present
E3: no signal from sensor 3 when danger is present
E: no signal from any of the three sensors when danger is present

$$\text{E} = \text{E1} \cap \text{E2} \cap \text{E3} \text{ (all must fail to give a signal for no signal from any)} \tag{3.13}$$

$$\text{P (E)} = \text{P (E1)} \cdot \text{P(E2)} \cdot \text{P(E3)} = (10^{-1})^3 = 10^{-3} \tag{3.14}$$

3.3.3 Series parallel networks

For yet more complex systems, it is possible to have both series and parallel components present in a single system. Many process systems have parallel streams in parts of the system, to maintain production whilst maintenance work is carried out or when equipment failures in one of the streams occur. In order to evaluate the reliability of such systems, network reduction techniques must be used. The network reduction method below is suitable for relatively simple series-parallel nets. The reliability can be evaluated reducing the network sequentially by combining series and parallel branches of the reliability model, until a single equivalent element remains.

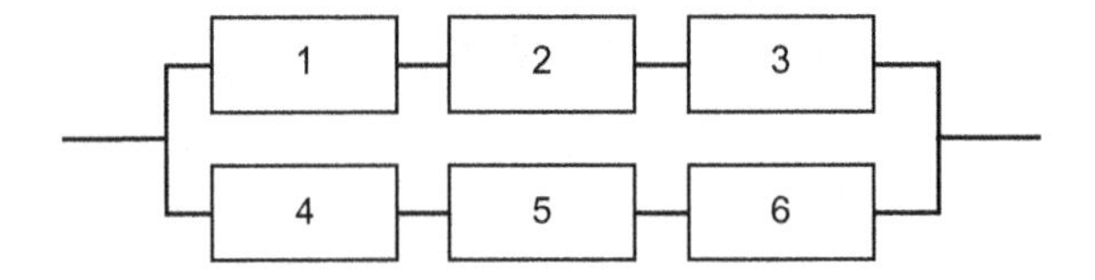

Figure 3.6 A simple series-parallel network.

For the first reduction combine the series elements (1, 2 and 3) and (3, 4 and 5) to give:

$$R_7 = R_1 \cdot R_2 \cdot R_3 \qquad R_8 = R_4\, R_5\, R_6 \tag{3.15}$$

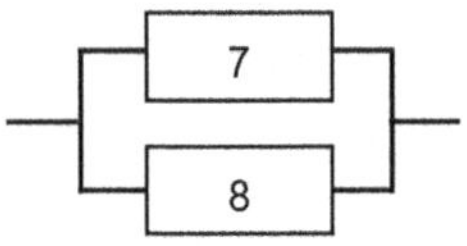

Figure 3.7 Initial reduction of the series elements of the network.

For the second reduction combine parallel elements 7 and 8 to give:

$$Q_9 = Q_7 \cdot Q_8 \tag{3.16}$$

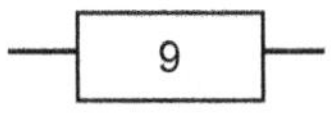

Figure 3.8 Final reduction of the parallel elements of the network.

The reliability of the system is therefore:

$$R_9 = (1 - Q_9) = (1 - [(1 - R_7)(1 - R_8)] \tag{3.17}$$

$$\text{or, } R_9 = (1 - [(1 - R_1\, R_2\, R_3)(1 - R_4\, R_5\, R_6)] \tag{3.18}$$

3.3.4 Partial redundancy

In many cases where redundant systems are installed, the redundancy is incomplete *i.e.* there is partial redundancy. Additional methods must be introduced to successfully reduce the network. The following network is an RBD for a system with a partially redundant sub-system called A.

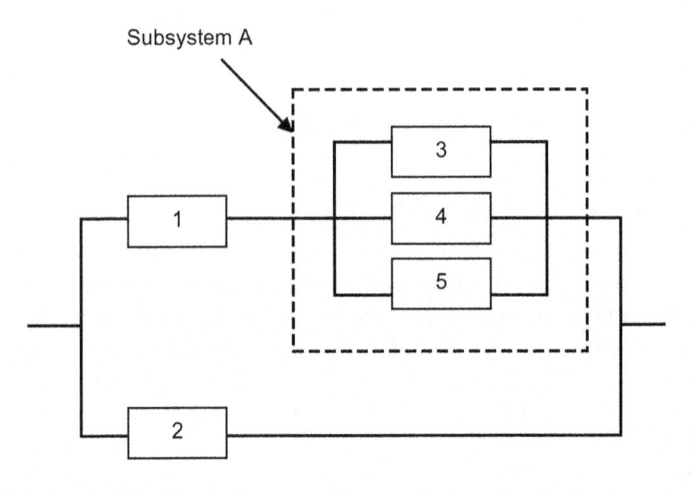

Figure 3.9 A reliability block diagram with a partially redundant sub-system.

In this case subsystem A is partially redundant –2 of the 3 components are required for this section to work. As an exercise, reduce this tree, assuming the reliability of each component to be identical and with $R = 0.9$ for each component and estimate the overall reliability of the system. For this example an additional method is required. Use the binomial distribution to work out the reliability of subsystem A. Then reduce the network using the same method as that used previously.

3.3.5 Standby redundant systems

It is frequently necessary for redundant systems to be installed in standby mode. Examples of standby redundant systems are often found in safety systems such as standby pumps and standby generators.

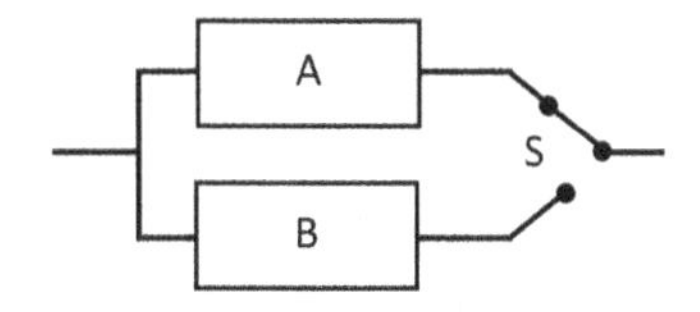

Figure 3.10 A standby redundant system.

In stand-by redundancy there are two additional considerations, namely:

- the duty of the stand by component; and
- the reliability of switching.

It is not difficult to see why the duty of the standby component is important. Consider the characteristics of a standby pump. In standby mode, the pump will not be operating, therefore wear is reduced, which extends the life of the standby pump. On the other hand, the pump may contain stationary fluids that could be corrosive and that could shorten its life, depending on the materials and the corrosivity of the fluids. The failure mode of, and therefore the maintenance work on the standby pump is likely to be quite different from that required for the operating pump. It follows that the reliability of pumps in standby mode is likely to be different from operating pumps.

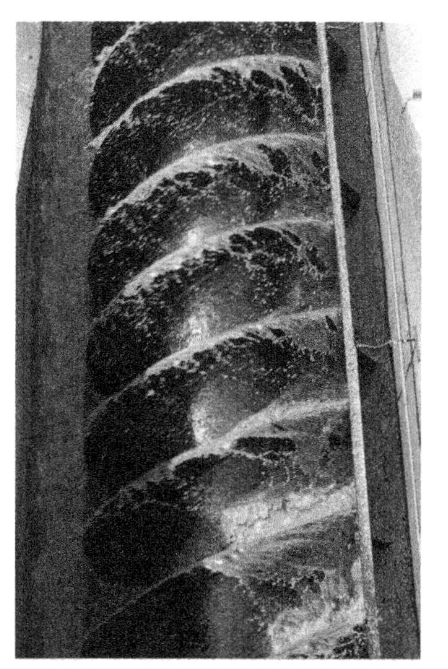

Credit: istockphoto © sauletas

Switching may be automatic or manual. The reliability of the switching system must therefore be included in the reliability analysis. For manual switching the reliability is dependent on human reliability – the ability to recognise a failure and to act quickly.

The probability of system failure for perfect switches is the probability that A fails **and** the probability that B is failed given that A has failed.

$$P(F) = P(\bar{A}) \cdot P(\bar{B}|\bar{B}) \tag{3.19}$$

If the failures of A and B are totally independent of each other, then the probability of failure is given by:

$$P(F) = P(\bar{A}) \cdot P(\bar{B}) \tag{3.20}$$

With an imperfect switch, there is a probability that the switch will fail when transferring from system A to system B. Let S represent successful switch-over and F system failure and use conditional probability to compute P(F) as follows:

The probability of failure of the switched system is:

$$P(F) = P(F|S) \cdot P(S) + P(F|\bar{S}) \cdot P(\bar{S}) \tag{3.21}$$

For the situation where the switch is assumed to be working, the probability of failure is given by

$$P(F|S) = P(\bar{A}) \cdot P(\bar{B}) \tag{3.22}$$

i.e. both A and B must be in the failed state if the switch worked.

For the situation where the switch is assumed to be failed, the probability of failure is given by:

$$P(F|\bar{S}) = P(\bar{A}) \tag{3.23}$$

If the switch is in the failed state it doesn't matter whether B is working or not, the system cannot operate. The conditional probabilities in equations 2 and 3 are then factored to take account of the probabilistic nature of the switching mechanism i.e. the fact that switch success or switch failure is not certain.

$$P(F) = P(\bar{A}) \cdot P(\bar{B}) \cdot P(S) + P(\bar{A}) \cdot P(\bar{S}) \tag{3.24}$$

3.3.6 Evaluation of complex systems

The above techniques are limited to simple systems, but many systems have additional complex logic that requires more sophisticated methods of analysis. Consider the following reliability block diagram for example:

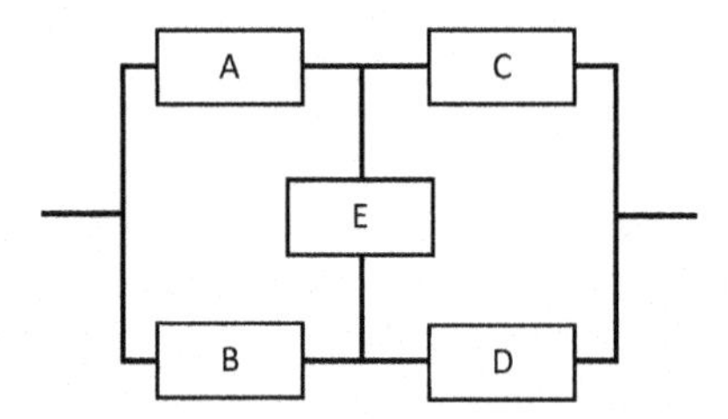

Figure 3.11 Reliability block diagram showing a complex system.

It is clear, by simple inspection, that this network cannot be redrawn so it conforms to the simple series – parallel networks above. This type of net requires different techniques. We will briefly consider the conditional probability method; the cut set method; and the event tree method.

In the conditional probability method, assumptions are made about the state of one of the components of the system, so that a simpler system can be assessed. That is, one of the components is assumed first to be in the failed state and then in the working state and this approach allows the complex cross over type of network to be reduced to a series-parallel type of net. The series-parallel nets are derived by representing a working component in the network as a short circuit. Similarly, when a component is assumed failed, it is represented as an open circuit on the network. It does not matter which component is selected, although the reduction may be simpler with one specific component. If we selected component E for the conditional assessment, we can reduce the system sequentially into sub system structures connected in series and parallel and then recombine as illustrated below. The failure probability of the overall system can then be calculated from:

$$P(F) = P(F|E) \cdot P(E) + P(F|\bar{E}) \cdot P(\bar{E}) \tag{3.25}$$

Evaluate the circuit on the left to give P(F|E) and on the right to give $P(F|\bar{E})$. Then combine the two to obtain P(F) using:

$$P(F) = P(F|E) \cdot P(E) + P(F|\bar{E}) \cdot P(\bar{E}) \tag{3.26}$$

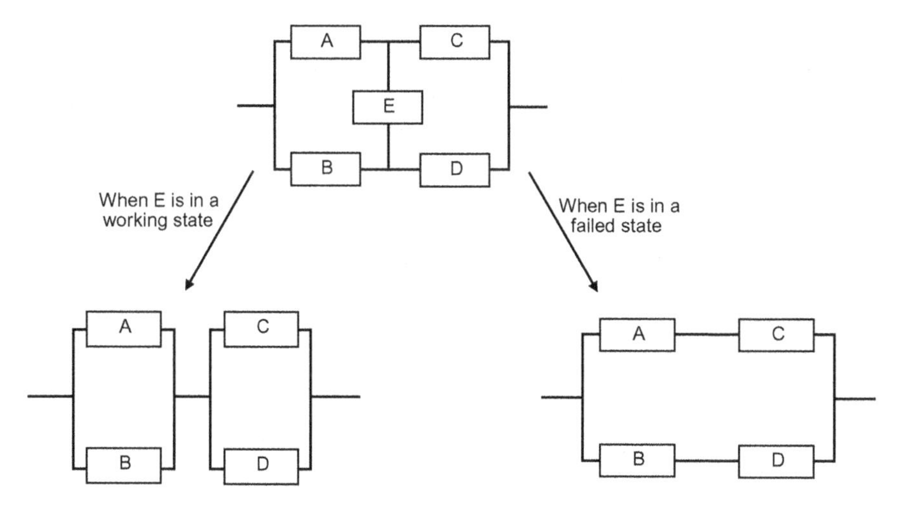

Figure 3.12 Evaluating a complex system using conditional probability.

3.3.7 Exercise

If p is the probability of failure for each of the subsystems A – E, show that the probability of failure of the system is given by

$$P(F) = 2p^2 + 2p^3 - 5p^4 + 2p^5 \tag{3.27}$$

The cut set method is a powerful method which is easily programmed on a computer for efficient solution of general networks. The cut sets are directly related to modes of failure; "a cut set is the set of system components which, when failed, causes the system to fail". In terms of a network of unit treatment processes linked as a treatment system, this definition can be interpreted to mean the set of components that must fail in order to disrupt all the paths between input and output.

The minimal cutset is a set of system components which, when failed, cause system failure **but** when only one component of the set is returned to the working state, the whole system returns to the working state.

We can use the same network as that used for the conditional probability method for comparison purposes. The method of determining the cut sets is to draw lines through the components of the network, as shown below, to interrupt all paths through the network. Each line cuts through the components which make up the cut set. In this example the cut sets are in fact minimal cut sets.

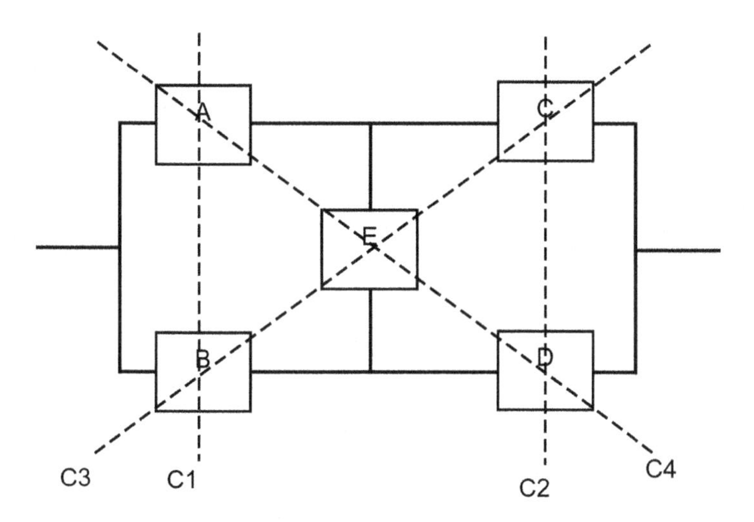

Figure 3.13 Evaluating a complex system using cut sets.

The cut sets found by inspection are:

C1	AB
C2	CD
C3	BEC
C4	AED

Thus there are 4 possible ways for the system to fail and the probability of failure is given by:

$$P(F) = P(C1 \cup C2 \cup C3 \cup C4) \quad (3.28)$$

The system fails when anyone of the following conditions is true:

A and B fail	cutset 1
C and D fail	cutset 2
B and E and C fail	cutset 3
A and E and D fail	cutset 4

This is equivalent to four subsystems in series as shown below:

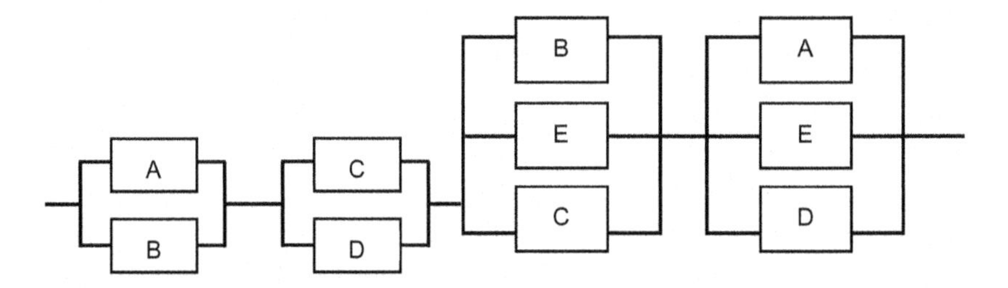

Figure 3.14 Cut sets shown as four subsystems in series.

The cut sets are effectively in series, but the series-parallel method of evaluation cannot be used since the same components appear in more than one cut set, e.g. D appears in cut 2 and cut 4. However, the concept of UNION does apply but it is a long and somewhat tedious approach.

3.3.8 Approximate evaluation of cut sets

The application of the Union method for evaluation of cut sets is time consuming. However the following approximate method is rapid and generally accurate enough. In general:

$$P(F) = P(C1 \cup C2 \cup C3 \cup C4) \quad (3.29)$$

To a 1st approximation:

$$P(F) = P(C1) + P(C2) + P(C3) + P(C4) \quad (3.30)$$

For i cut sets

$$P(F) = \Sigma P(C_i) \quad (3.31)$$

Table 3.4 Cut set modes and probabilities.

Cuts	Failure Mode	Probability	
1	A and B fail	$P(\bar{A}) \cdot P(\bar{B})$	p^2
2	C and D fail	$P(\bar{C}) \cdot P(\bar{D})$	p^2
3	A, D and E fail	$P(\bar{A}) \cdot P(\bar{D}) \cdot P(\bar{E})$	p^3
4	B, C and E fail	$P(\bar{B}) \cdot P(\bar{C}) \cdot P(\bar{E})$	p^3

Hence:

$$P(F) = P(C1) + P(C2) + P(C3) + P(C4) \quad (3.32)$$

$$P(F) = P(A) \cdot P(B) + P(\bar{C}) \cdot P(D) + P(A) \cdot P(D) \cdot P(E) + P(B) \cdot P(\bar{C}) \cdot P(E) \quad (3.33)$$

When all probabilities are the same, p, this expression reduces to:

$$P(F) = 2p^2 + 2p^3 \quad (3.34)$$

It is instructive to compare values between the rigorous and approximate solutions. The accurate method gives:

$$P(F) = 2p^2 + 2p^3 - 5p^4 + 2p^5 \quad (3.35)$$

and when $p = 0.01$ P(F) = 0.00020195

The approximate method gives:

$$P(F) = 2p^2 + 2p^3$$

and when $p = 0.01$ P(F) = 0.000202.

The 1st term gives 0.000200 and represents an upper bound on reliability. The 1st and 2nd terms give 0.000202 which represents a lower bound on reliability. The accurate value using all terms gives 0.00020195. These results illustrate the theoretical application of cut sets, but can rarely be applied at this level of sophistication for water and wastewater utilities. We now explore process risk analysis tools that utilise data more akin to those available to risk analysts in the sector and specifically the application of event trees to the analysis of microbiological risk from pathogens.

3.4 MICROBIOLOGICAL RISK ASSESSMENT

Microbiological risk assessment (MRA) is the risk analysis technique used to predict and model the risks to human and animal health from pathogens in the environment. It can be used to answer questions such as *'by how much will public health be jeopardised if a major treatment process fails'*?. It does this by estimating the probability of human exposure to an infectious dose of a pathogenic agent through environmental transmission routes. This estimated dose is then compared with doses known to cause harm. MRA has become a discipline in its own right and is widely used in veterinary science, medical microbiology and environmental microbiology. As with all risks involving hazardous agents that leave contained, or semi-contained systems, one must first evaluate the likely loss of containment, or evasion of a contained system, and then the onward exposure in the wider environment. For microbiological risks and water treatment, there is the issue of the potential entry of pathogens into water treatment systems from the catchment and the onward pathogen challenge to unit processes and their treatment effectiveness.

Credit: istockphoto © FrankvandenBergh

Microbial risks are managed through the implementation of barriers to exposure – think of the use of surgical gloves in hospitals, for example, as a primary barrier; or the use of antimicrobial gels for hand washing. Protective barriers, once evaded by pathogens, allow onward exposure. Of key importance in MRA then is the initial identification of protective barriers. There are two types of barrier: pathway barriers that control how much infectivity receptors are exposed to; and biomedical barriers that control how infectious the agent is to the receptor once exposure has occurred.

Examples of pathway barriers for drinking water systems include the drinking water treatment processes (filtration, coagulation) and the hydrogeological sub-strata above aquifers that filter out microorganisms.

Examples of the biomedical barriers include the species barrier in the case of human exposure to the infective bovine spongiform encalopathy agent; acquired protective immunity following *C. parvum* exposure; and natural gut microbiota following salmonella expsoure.

Concentrations of pathogens fluctuate widely in the aquatic environment, by orders of magnitude, and so we typically refer to base 10 logarithmic level of exposure or removal ($\log_{10}$). Ineffective removal of *Cryptosporidium* oocysts, during water treatment, is a well-documented cause of waterborne outbreaks of cryptosporidiosis. There is variation within the human population in their susceptibility to pathogens, reflecting pathogen strain and acquired protective immunity. The importance of accommodating the natural variation is a central theme within MRA and in terms of treatment processes, by-pass at operational scale is of fundamental importance.

One problem facing utility managers is how to determine whether pathogen removal by drinking water treatment has been improved by, say, 10-fold. This is complicated by the variation in the pathogen removal ratios by the unit processes of drinking water treatment (Table 3.6). Remember also that for unit processes in series, the system probability is only as strong as its weakest link.

Table 3.5 'Source', 'pathway' and 'receptor' terms applied to *Cryptosporidium* oocysts in drinking water (after Gale & Stanfield, 2000).

Term	Data	Variation	Uncertainty
Source	Oocyst concentrations in raw water	Natural fluctuations in oocyst concentrations, e.g., after heavy rainfall	Monitoring protocol missing high count spikes and recovery efficiency of analytical method contribute to underestimating net oocyst loading
Pathway	Removal of oocysts from raw water by drinking water treatment	Fluctuations in removal efficiency between and within individual treatment works	Monitoring protocol may cause net overestimation of net removal efficiency
Receptor	Volume of tap water ingested daily by individual consumers	Individual habits, age groups, seasonal effects, amount imbided after boiling	Method of data collection, e.g. self-assessment

A theoretical example of fluctuations in pathogen removal efficiency is shown in Figure 3.15. Here, the removal ratio for the whole system, comprised of individual unit processes, varies between 1-log and 5-logs. The average log-removal ($\mu_{\log}$) is 2.7-logs and the geometric mean (or median) removal is calculated as $10^{2.7}$ (a 500-fold removal). However, the net (or arithmetic mean) removal is much less, and depends on the variation and in particular, the frequency of poor removal efficiencies, especially on 'bad days', indicated by the bold arrows in Figure 3.15. 'Bad days' may be effected by variable filter performance, or sub-optimal filter reliability inferring the need for filter maintenance and backwash, say. Thus, for the dashed line in Figure 3.15, the net removal ratio is 1.96-logs (91-fold). For the solid line, the net removal ratio is 1.78-logs (61-fold). The solid line has a greater variance, and in particular more poor removal frequencies balanced by more good removal frequencies. However, the 'bad days' serve to decrease the arithmetic mean (or net) removal. In terms of net removal, the 'bad days', where the pathogen removal is poor, cannot be compensated by 'good days' and thus dictate system performance. This is why preventative maintenance is so critical for water and wastewater treatment unit processes.

Credit: istockphoto © esvetleishaya

Table 3.6 Waterborne pathogens and their significance in water supplies (after World Health Organisation, 2007). In this table, the waterborne transmission of the pathogens listed has been confirmed by epidemiological studies and case histories. Part of the demonstration of pathogenicity involves reproducing the disease in suitable hosts. Experimental studies in which volunteers are exposed to known numbers of pathogens provide relative information. As most studies are done with healthy adult volunteers, such data are applicable to only a part of the exposed population, and extrapolation to more sensitive groups is an issue that should be studied in more detail.

Pathogen[a]	Health Significance[b]	Persistence in Water Supplies[c]	Resistance to Chlorine[d]	Relative Infectivity[e]	Important Animal source
Bacteria					
Burkholderia pseudomallei	High	May multiply	Low	Low	No
Campylobacter jejuni, C. coli	High	Moderate	Low	Moderate	Yes
Escherichia coli – Pathogenic[f]	High	Moderate	Low	Low	Yes
E. coli – Enterohaemorrhagic	High	Moderate	Low	High	Yes
Legionella spp.	High	May multiply	Low	Moderate	No
Non-tuberculous mycobacteria	Low	May multiply	High	Low	No
Pseudomonas aeruginosa[g]	Moderate	May multiply	Moderate	Low	No
Salmonella typhi	High	Moderate	Low	Low	No
Other salmonellae	High	May multiply	Low	Low	Yes
Shigella spp.	High	Short	Low	High	No
Vibrio cholerae	High	Short to long[h]	Low	Low	No
Yersinia enterocolitica	High	Long	Low	Low	Yes
Viruses					
Adenoviruses	High	Long	Moderate	High	No
Enteroviruses	High	Long	Moderate	High	No
Astroviruses	High	Long	Moderate	High	No
Hepatitis A viruses	High	Long	Moderate	High	No
Hepatitis E viruses	High	Long	Moderate	High	Potentially
Noroviruses	High	Long	Moderate	High	Potentially
Sapoviruses	High	Long	Moderate	High	Potentially
Rotavirus	High	Long	Moderate	High	No
Protozoa					
Acanthamoeba spp.	High	May multiply	Low	High	No
Cryptosporidium parvum	High	Long	High	High	Yes
Cyclospora cayetanensis	High	Long	High	High	No
Entamoeba histolytica	High	Moderate	High	High	No
Giardia intestinalis	High	Moderate	High	High	Yes
Naegleria fowleri	High	May multiply[i]	Low	Moderate	No
Toxoplasma gondii	High	Long	High	High	Yes
Helminths					
Dracunculus medinensis	High	Moderate	Moderate	High	No
Schistosoma spp.	High	Short	Moderate	High	Yes

[a]Pathogens for which there is some evidence of health significance related to their occurrence in drinking-water supplies; [b]Health significance relates to the severity of impact, including association with outbreaks; [c]Detection period for infective stage in water at 20°C: short, up to 1 week; moderate, 1 week to 1 month; long, over 1 month; [d]When the infective stage is freely suspended in water treated at conventional doses and contact times and pH between 7 and 8. Low means that 99% inactivation at 20°C generally in <1 minute, moderate 1–30 minutes and high >30 minutes; [e]From experiments with human volunteers, from epidemiological evidence and animal studies. High means infective doses can be $1–10^2$ organisms or particles, moderate $10^2–10^4$ and low $>10^4$; [f]Includes enteropathogenic, enterotoxigenic and enteroinvasive; [g]Main route of infection is by skin contact, but can infect immunosuppressed or cancer patients orally; [h]*Vibrio cholerae* may persist for long periods in association with copepods and other aquatic organisms; [i]In warm water.

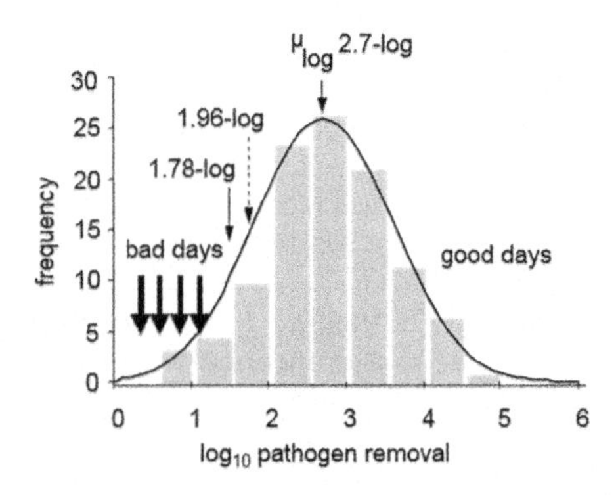

Figure 3.15 A theoretical model for variation in log pathogen removal by a drinking water treatment plant (adapted from Gale & Stanfield, 2000). Good days do not compensate for bad days.

Monitoring programmes to estimate pathogen removals for risk assessment calculations should therefore be designed to detect the 'bad days'. The distribution of removal ratios in Figure 3.15 represents those which would be obtained on taking single samples of raw and treated water. It is apparent that 50% of those measurements would suggest a >500-fold removal when in fact the net removal is only 61-fold (for the solid line). This illustrates the importance of understanding the variation. The same principles apply to designing experiments to determine the net pathogen destruction by sludge treatment processes, for example.

3.4.1 Case study – crop exposures to pathogens in sewage sludge

Quantitative events trees have been widely used in MRA to estimate log removals of pathogens during unit process treatment. They must be used with caution as they embody numerous assumptions on the assumed homogeneity of systems and averaged behaviours, removals and exposures. Nevertheless, they have proved useful for illustrating key processes for removal, the importance of eliminating 'bypass' in systems and thus for identifying critical control points.

Here is an example related to the land application of biosolids. In the UK, in 1996/97, some 520,000 tonnes dry solids (tds) of treated sewage sludge were disposed of by application to agricultural land per annum. Concern has historically been raised about the potential impact of this activity on the food chain and the microbiological risks that might be posed to food crops. Gale and colleagues (2001, 2002, 2003) have used simple event trees to explore the magnitude of these and similar risks. Figure 3.16 is an event tree developed to model the partitioning of salmonella levels from raw sewage into raw sludge at a sewage treatment works. A further event tree then models the transmission of salmonellas in the raw sludge to root crops, such as potatoes, at the point of harvest (Figure 3.17).

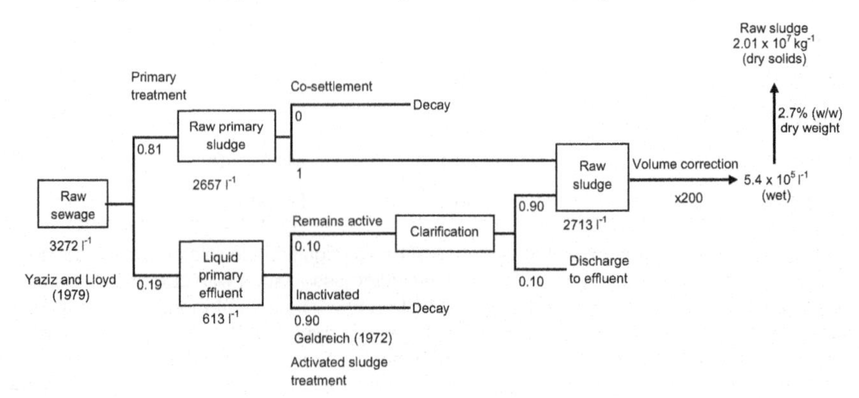

Figure 3.16 Event tree for partitioning of Salmonella into raw sewage sludge (redrawn from Gale, 2003).

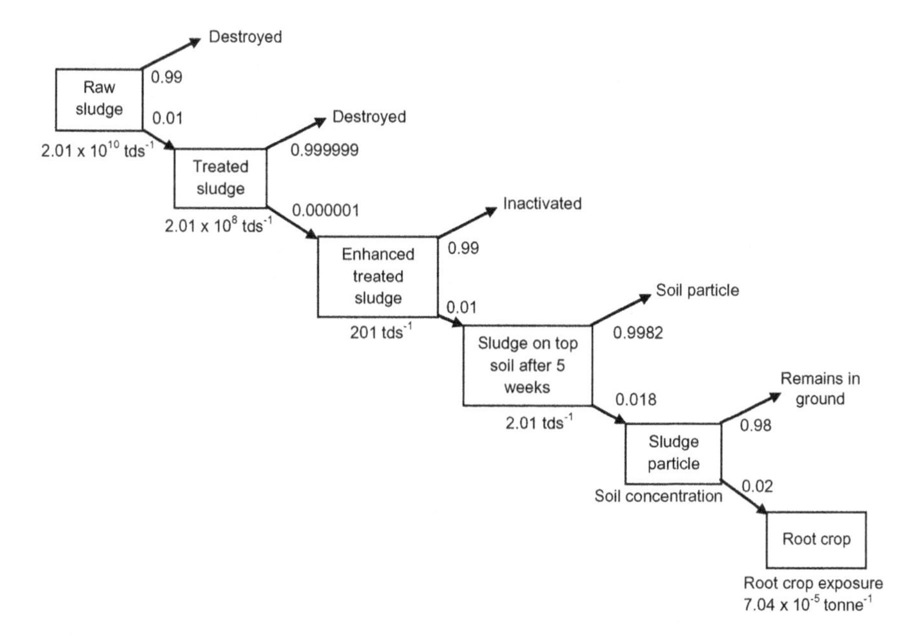

Figure 3.17 Event tree for transmission of salmonella in raw sewage sludge to potatoes (redrawn from Gale, 2003).

3.4.2 The critical importance of bypass

What happens when otherwise effective treatment is compromised by pathogens (or chemicals) bypassing control (pathway) barriers? In short, treated product becomes contaminated with raw material. Event trees are useful for illustrating this phenomenon (Figure 3.18). Under bypass conditions, the failure of seemingly minor routes have dramatic impacts of the effectiveness of risk management controls, in this case drinking water treatment. This severely limits the effectiveness of HACCP procedures put in place for the major route and we can observe (Gale, 2003) that the minor routes become more important as the major routes are tightened up (Table 3.6). Indeed, if bypass continues to be observed and remains unaddressed, there comes a point when a 10-fold improvement in water quality through one route has only a 0.1% impact on the UK picture (Table 3.7).

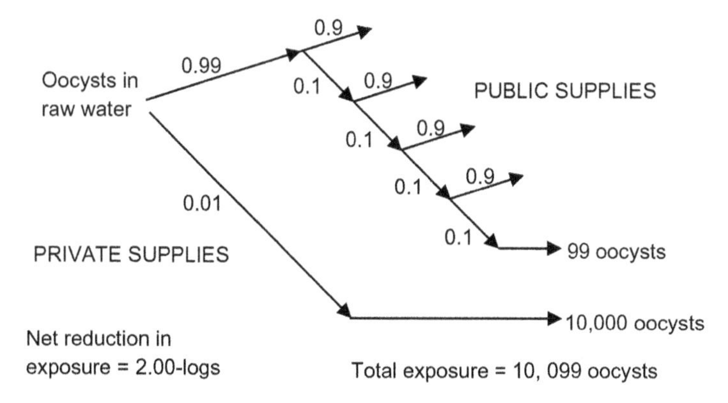

Figure 3.18 Impact of bypass on treatment efficiency illustrated for the consequences of *C. Parvum* bypassing treatment in untreated private drinking water supplies.

We have seen in this unit, through various mathematical expressions and worked examples that being conscious of the dramatic impact of 'bypass' is an important requirement for water quality managers. Vigilance is required in moments of changing conditions (e.g. enhanced run-off, flooding) to ensure that system integrity is maintained.

Table 3.7 Bypass and the law of diminishing returns.

Major Route 99%	Minor Route 1%	Net Reduction on Population
1-log	No treatment	0.96-log
2-log	No treatment	1.70-log
3-log	No treatment	1.96-log
4-log	No treatment	1.99-log
5-log	No treatment	2.00-log
6-log	No treatment	2.00-log

3.5 SUMMARY AND SELF-ASSESSMENT QUESTIONS

In this unit, we have considered the context of process risk analysis and the tools and techniques that can be applied in the water and wastewater utility sector to assess risk and evaluate the reliability of linked unit treatment processes. In the next unit, we begin to look beyond the process boundary and consider other contributors to failure. Our discussion will extend beyond engineered systems to other sources of hazard and will include a discussion of human and organisation risk management. As the text progresses, we will begin to consider the broader context of improved decision-making within the utility sector – that is, the application and implementation of risk analysis and management for more credible and defensible decisions. This will take us well beyond probability and reliability theory into the environmental, management and social sciences.

SAQ 3.1 Consider an activated sludge treatment process. How do the terms 'reliability', 'availability' and 'maintainability' apply to this unit process?

SAQ 3.2 Consider the equation in Example 3.1. How might this equation be amended to account for the time associated with the preventative maintenance (PM) of unit processes?

3.6 FURTHER READING

Adkin A., Donaldson N. and Kelly L. (2013). A quantitative assessment of the prion risk associated with wastewater from carcase-handling facilities. *Risk Analysis*, **33**, 1212–1227.

American Institute of Chemical Engineers (1992). Guidelines for Hazard Evaluation Procedures, 2nd edn, American Institute of Chemical Engineers, New York, US.

Ander H. and Forss M. (2011). Microbiological Risk Assessment of the Water Reclamation Plant in Windhoek, Namibia, MSc thesis, Department of Civil and Environmental Engineering, Chalmers University of Technology, Sweden http://publications.lib.chalmers.se/records/fulltext/150138.pdf

Ashbolt N. (2004). Risk analysis of drinking water microbial contamination versus disinfection by-products (DBPs). *Toxicology*, **198**, 255–262.

Christodoulou S. E. (2011). Water network assessment and reliability analysis by use of survival analysis. *Water Resources Management*, **25**, 1229–1238.

Comer P. J. and Huntly P. J. (2003). TSE risk assessments: A decision support tool. *Statistical Methods in Medical Research*, **12**, 279–291.

Cyna B. (1997). Reliability analyses of water treatment plants. *Water Supply*, **15**, 65.

Debón A., Carrión A., Cabrera E. and Solano H. (2010). Comparing risk of failure models in water supply networks using ROC curves. *Reliability Engineering and System Safety*, **95**, 43–48.

Drinking Water Inspectorate (2005). A brief guide to water safety plans, October 2005, 12pp, available at: www.dwi.gov.uk/guidance/Guide%20to%20wsp.pdf.

Gale P., Young C., Stanfield G. and Oakes D. (1998). Development of a risk assessment for BSE in the aquatic environment. *J. Appl. Microbiology*, **84**, 467–477.

Gale P. and Stanfield G. (2000). *Cryptosporidium* during a simulated outbreak. *Journal of American Water Works Association*, **92**, 105–116.

Gale P. (2001). Developments in microbiological risk assessment for drinking water – a review. *J. Appl. Microbiology*, **91**, 191–205.
Gale P. and Stanfield G. (2001). Towards a quantitative risk assessment for BSE in sewage sludge. *J. Appl. Microbiology*, **91**, 563–569.
Gale P. (2002). Using risk assessment to identify future research requirements. *Journal American Water Works Association*, **94**, 30–38.
Gale P. (2003). Using event trees to quantify pathogen levels on root crops from land application of treated sewage sludge. *J. Appl. Microbiology*, **94**, 35–47.
Gómez Fernández J. F. and Crespo Márquez A. (2009). Framework for implementation of maintenance management in distribution network service providers. *Reliability Engineering and System Safety*, **94**, 1639–1649.
Haas C. N. et al. (1996). Assessing the risk posed by oocysts in drinking water. *Journal of American Water Works Association*, **88**, 131–136.
Hass C. N., Rose J. B. and Gerba C. P. (2014). Quantitative Microbial Risk Assessment, 2nd edn, Wiley, 440 pp.
Hellier K. (2000). Hazard analysis and critical control points for water supplies. *Presented at the 63rd Annual Water Industry Engineers and Operators' Conference*, Warrnambool, Australia, 6-7 September 2000, pp. 101–109.
Martínez-Rodríguez J. B., Montalvo I., Izquierdo J. and Pérez-García R. (2011). Reliability and tolerance comparison in water supply networks. *Water Resources Management*, **25**, 1437–1448.
Mena K. D., Gerba C. P., Haas C. N. and Rose J. B. (2003). Risk Assessment of waterborne Cosxackievirus. *J. Amer. Water Works Assoc.*, **95**, 122–131.
Pitblado R. and Turney R. (1996). Risk Assessment in the Process Industries, 2nd edn, Institution of Chemical Engineers, Rubgy, UK.
Rose J. B. and Gerba C. P. (1991). Use of risk assessment for development of microbial standards. *Water Science and Technology*, **24**, 29–34.
Rose J. B. and Masago Y. (2007). A toast to our health: Our journey toward safe water. *Water Science & Technology: Water Supply*, **7**, 41–48.
Signor R. and Ashbolt N. J. (2009). Comparing probabilistic microbial risk assessments for drinking water against daily rather than annualised infection probability targets. *Journal of Water & Health*, **7**, 535–543.
Schijven J. F. et al. (2011). QMRAspot: A tool for quantitative microbial risk assessment from surface water to potable water. *Water Research*, **45**, 5564–5576.
Smeets P. W. M. H. (2010). Stochastic Modelling of Drinking Water Treatment in Quantitative Microbial Risk Assessment. KWR and IWA Publishing, London, UK.
Todinov M. T. (2005). Reliability and Risk Models: Setting Reliability Requirements. John Wiley & Sons Ltd., Chichester, UK.
Taheriyoun M. and Moradinejad S. (2015). Reliability analysis of a wastewater treatment plant using fault tree analysis and Monte Carlo simulation. *Environmental Monitoring and Assessment*, **187**, 4186.
Tchorzewska-Cieslak B. K., Boryczko K. and Eid M. (2012). Failure scenarios in water supply system by means of fault tree analysis. *Advances in Safety, Reliability and Risk Management*, Taylor & Francis, pp. 2492–2499.
Torres J. M., Brumbelow K. and Guikema S. D. (2009). Risk classification and uncertainty propagation for virtual water distribution systems. *Reliability Engineering and System Safety*, **94**, 1259–1273.
USEPA (2014). Microbiological Risk Assessment (MRA). Tools, Methods, and Approaches for Water Media. Office of Water, USEPA, Washington DC.
Wang Y. and Au S. -K. (2009). Spatial distribution of water supply reliability and critical links of water supply to crucial water consumers under an earthquake. *Reliability Engineering and System Safety*, **94**, 534–541.
World Health Organisation (2007). Waterborne pathogens at: http://www.who.int/water_sanitation_health/gdwqrevision/watborpath/en/

Unit 4

Assessing risks beyond the process boundary

HOW MUCH CONTROL OVER A HAZARD AND WHICH REGULATORY APPROACH?

Diffuse pollution is harder to manage than point source pollution. Frequently, it requires changing the behaviours of those that undertake hazardous activities that result in more dispersed releases within catchments. This raises questions about how these activities are best regulated and by whom.

For water utilities, direct regulation has been regarded as providing stability for investment planning, and necessary control for high risk activities. More broadly, beyond the process boundary, greater flexibility for those seeking to voluntarily try out new approaches to controlling hazards might yield greater benefits. For catchment stakeholders going beyond compliance, 'earned recognition' might be welcomed, accommodated by good relationships with regulators. Regulation that accommodates local decision making and clear accountabilities for risk management is necessary for flood management.

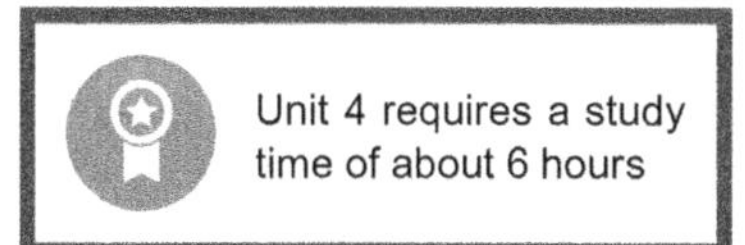

4.1 INTRODUCTION

Our consideration of risk assessment and management now moves beyond unit treatment processes (engineering risk and reliability) to risks up, and downstream, of a water or wastewater treatment plant (Table 4.1), and to their identification, assessment and management.

Examination of risks in this category, and of tools for their assessment and management, is necessary because: (i) certain upstream hazards associated with activities within catchments pose a risk to the raw water supply and, therefore, to the reliability of unit processes for water treatment and ultimately to the quality of the potable supply; (ii) trade effluent discharges to the foul sewer may impact adversely on the influent quality of wastewaters at a sewage works, impacting on treatment efficacy and, therefore, the quality of discharged effluent, post-treatment; and (iii) many downstream risks, subsequent to the distribution of treated water, or to the discharge of treated wastewaters, are difficult to manage once process control over them has been lost at the works.

Critical risks in this category include (a) chemical and microbiological risks in catchments affecting raw water quality upstream of water treatment processes; for example the shedding of pathogen loads by farmed animals or the loss of sediment load from farmland affecting river water quality; (b) pathogenic risks within water distribution networks, downstream of water treatment works, managed by the maintenance of a chlorine residual; (c) risks from lead exposure in the public supply managed through phosphate addition or progressive pipe replacement programmes; (d) risks to receiving waterbodies, post treatment, from wastewaters containing difficult to assimilate BOD, persistent trace nutrients or bioaccumulative hazardous compounds.

Prior knowledge of the basic concepts of water quality management and the component stages of water supply and wastewater treatment system is assumed in this Unit (see further reading). For water supply, this encompasses distinct points in the water supply cycle: at resource, during treatment, during distribution and within the household plumbing system – that is, from catchment to tap. For wastewaters, this includes knowledge of the processes from sewage collection (including runoff), industrial wastewater discharge to sewer, through treatment processes to effluent discharge, post treatment.

Beyond the unit process boundaries, environmental systems vary tremendously in the types of risk problem they pose and they harbour greater uncertainty because there is less control over hazardous activities in open systems (Figure 4.1). Infrastructures are frequently buried and not easy to monitor, or activities that impact on water quality may be in the control of other stakeholders; the pesticide spraying of crops, or the farming of sediment-shedding crops, for example. As a result of a whole series of catchment activity with point- or diffuse source origins, the water resources available to utilities may have inherent and undesirable constituents that require removal (e.g. natural organic matter; iron, manganese; contamination from historically polluted land, excess phosphate). Within catchments, a host of land uses (Table 4.1) and land use changes, (aforestation, urbanisation, agricultural activity), pose hazards to the quality of raw waters for water supply. These drivers may then be exacerbated (or mitigated) by the growing impacts of climate or demographic change.

Credit: istockphoto © nu_andrei

Table 4.1 Anthropogenic activities affecting river systems (after Gray, 1999).

Supra-catchment effects
Acid deposition
Inter-basin transfers
Climate change impacts
Catchment land-use change
Afforestation and deforestation
Urbanisation
Agricultural development
Land drainage/flood protection
Groundwater remediation/natural attenuation
Corridor engineering
Removal or replacement of riparian vegetation
Flow regulation – dams, channelization, weirs etc.
Dredging and mining
Instream impacts
Organic and inorganic pollution
Thermal pollution
Abstraction
Navigation
Exploitation of native species

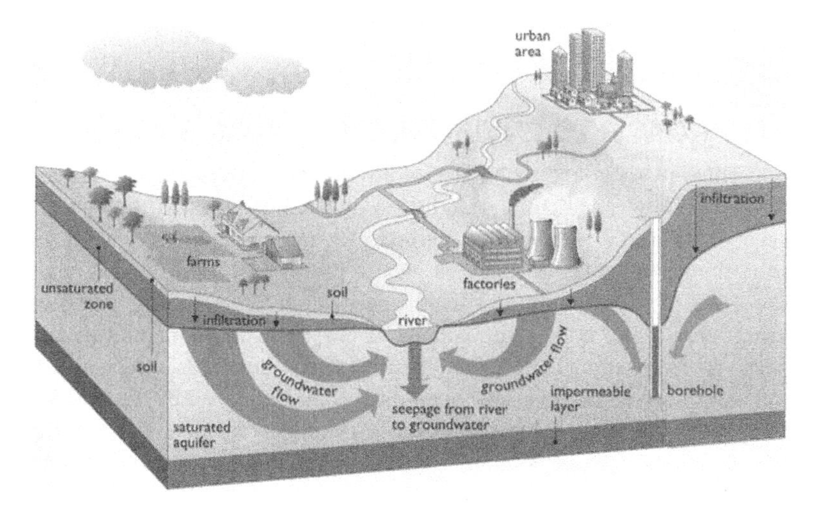

Figure 4.1 Beyond the process boundary. Open environmental systems such as catchments, contain many hazards and uncertainties. Their management requires a mixed strategy of consent compliance, land use management and behavioural change (© Natural Environment Research Council).

In open environmental systems, pollution is usually diffuse and risks best managed by inducing behavioural change, (e.g. adopting buffer strips), implementing codes of good industrial practice, (e.g. on biosolids application to land; on the use of sustainable urban drainage systems, on oil storage and bunding) or by adopting controls that limit hazardous activities at source in the catchment, (e.g. stipulating source protection controls to prevent groundwater pollution, enforcing protected catchments or establishing nitrate vulnerable zones). In Europe, the Water Framework Directive, a major piece of environmental legislation, requires the identification and prioritisation of hazards within 'river basin districts' for the purposes of improving the quality of water bodies to good ecological status.

For engineered structures outside the conventional treatment chain (those linked unit processes used for water and wastewater treatment) we are concerned with the progressive deterioration of structures, such as storage reservoirs or stormwater tanks, over time and with the likelihood and consequences of a catastrophic failure of these structures or of the flooding events that such failures might impart. With underground collection and distribution systems, attention turns to pipe failures, sewer collapses, mains bursts, pressure fluctuations, high flow situations and unsatisfactory combined storm overflow (CSO) discharges from networked systems.

For supply at the tap, we are principally concerned with potential risks to public health from in-home plumbing systems, (e.g. lead), or from chemicals and pathogens that may evade conventional treatment (e.g. *Cryptosporidium*) or that reach the tap because of some upstream loss in system performance. Here, our interest as risk assessors is in the likelihood of customers being exposed to excessive doses of chemicals, or pathogens, that may result in illness and therefore in the setting and maintenance of risk-based quality standards that protect against these adverse outcomes.

For discharged treated effluent, we are concerned about the likelihood of human and ecological impacts from discharges containing hazardous components, such as excess ammonia, organic matter that expresses a high biochemical oxygen demand (BOD) or specific contaminants, such as endocrine disrupting chemicals, for example.

Given this breadth of issues, the aim of this unit is to introduce a selection of tools and techniques used for evaluating these operational risks that exist beyond the unit process boundary. The unit objectives are to:

- extend risk assessment and management principles beyond unit processes;
- introduce the principles, frameworks and tools for environmental risk management;
- introduce 'source', 'pathway', 'receptor' concepts;

SECURING WATER QUALITY IN GERMANY

The German water supply sector is characterised by its high technical and managerial standard developed in an environment of self-regulation. The multiple barrier principle is firmly established in the DVGW technical standards series W110 to W135 for e.g. well construction and groundwater monitoring so that the risk of water contamination during abstraction is minimised. The DVGW series W200 to W296 sets out the principles of the second barrier in water treatment. The individual standards have internal barriers to minimise the risk of pollution further. The third barrier is effective disinfection and the DVGW technical standard and also the 'Trinkwasserverordnung' (Drinking Water Ordinance), based on the EU Drinking Water Directive, prescribe reliable disinfection procedures. The fourth barrier to protect safe drinking water is the prevention of secondary contamination in the distribution network.

- apply risk assessment from catchment to tap; and
- introduce human factors in risk management.

4.2 APPROACH TO ENVIRONMENTAL RISK MANAGEMENT

A wide range of tools and techniques are used for undertaking operational risk assessments. They range from straightforward examinations of the connectivity between the source of a hazard and a receptor, (an asset, environmental feature or living thing that we value and thus seek to protect), to sophisticated numerical packages for conducting probabilistic analysis on high quality data sets. In practice, many risk problems are addressed initially using a qualitative analysis. Complex environmental issues with significant consequences invariably require a combination of qualitative and quantitative analysis, usually because certain aspects of the system are better described, with supporting data, than others are.

It is essential at the outset of any risk assessment to be clear about the context, purpose and decision that the risk assessment is informing, or put another way, to very clearly understand *'the risk of what to whom'*. The context of the decision is critical and will be returned to throughout the analysis. The risk analysis should be logical and systematic and the risk assessor clear about how the output of the risk assessment will be used alongside other factors, (social, technological, management, economic) to inform a decision of how best to manage the risk, providing the significance of the risk requires this. These aspects should be considered in a formal problem definition stage (Figure 4.2).

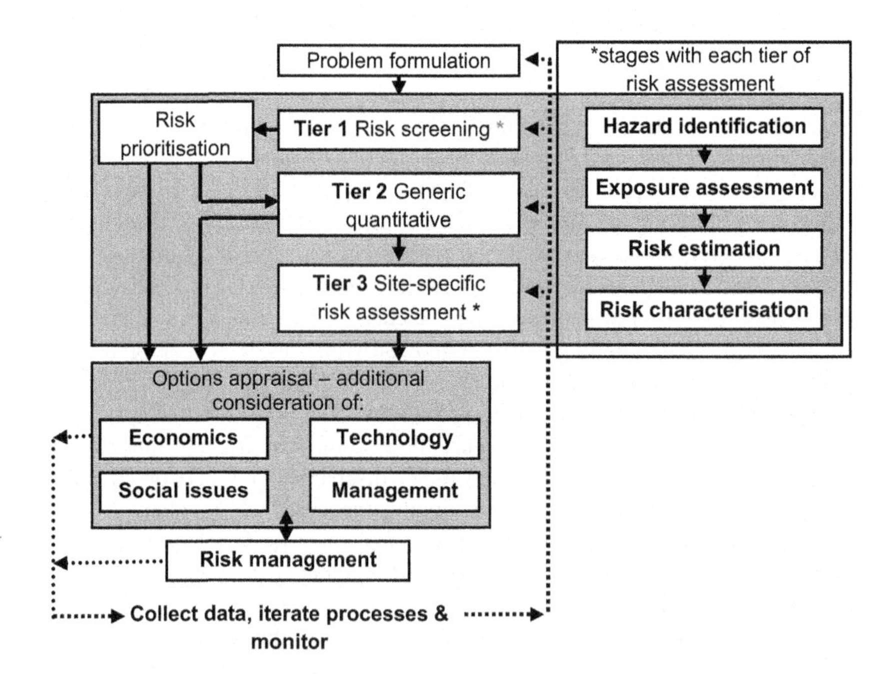

Figure 4.2 One approach to risk management – a generic risk management framework for environmental risks.

A tiered approach to risk assessment (Figure 4.2) allows for risk screening, prioritisation and a qualitative treatment in advance of quantification. Because there is gross uncertainty in open systems, particularly in the assessment of environmental exposures and impacts, resources should be targeted to where risks or uncertainties are high and where the costs of the risk assessment are likely to be justified by the benefits of risk management. A 'risk screening' approach is used (Tier 1; Figure 4.2) to determine key risks and priorities in advance of a more detailed analysis. If the decision on risk significance and risk management cannot be made using this approach, then more detailed tools are used focusing on the key risks prioritised during the screening stage (Tiers 2 and 3).

A primary consideration for the risk analyst is the type of risk being assessed – its character, or the way the risk is expressed. Knowing this provides essential information about which tools and techniques are best suited to its assessment. The risk analyst may be concerned with:

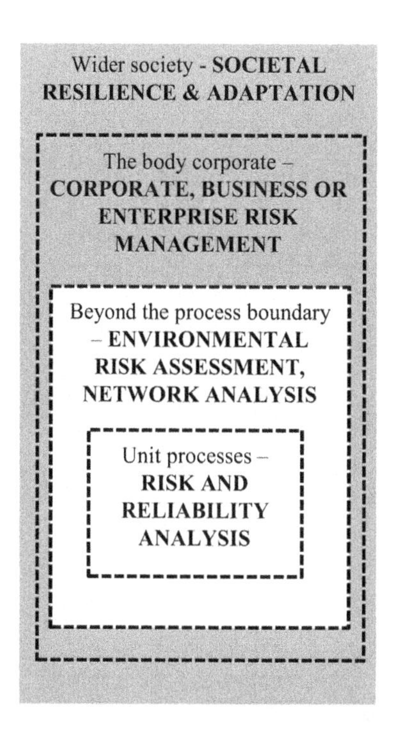

- the risk of an **initiating event**, or combination of events occurring that subsequently results in a **release to the environment**, (e.g. the over-topping of a coastal flood defence by large volumes of sea water; the release of firewater to site drainage systems from the over-topping or failure of an on-site collection tank; a process plant failure; the breakthrough of a filter press; a reservoir collapse; a CSO discharge); or with
- the risk of **exposure** to an environmental receptor following an initiating event that results in a release, (e.g. derogation in the quality of a drinking water supply that has become affected by a landfill leachate plume; the dispersion of contaminants in surface water downstream from consent failure; the exposure of a biological treatment process (e.g. an activated sludge plant) to shock loads of industrial cyanide); or with
- the risk of **harm** resulting from exposure, (e.g. risks to humans as a result of exposure to toxic/asphyxiant gases or odours; physical damage to stormwater tanks following the entry of flood waters; harm to a wetland following over-abstraction or periods of drought).

Selecting the right tool, in the right circumstance and for the right purpose requires:

(i) selecting the appropriate approach to the risk assessment, by reference to the type of risk above; and
(ii) selecting the appropriate level (tier) of sophistication for the tool as needs, complexities, priorities and data allow (Figure 4.2).

It is helpful to develop an early view on how the risks are likely to be managed. The management of hazards at source is usually preferred over the expensive retrofitting of process plant and so hazards such as contaminants in catchments or further upstream in consumer products are ideally managed by (i) replacement activities; (ii) managing behaviours in catchments, (e.g. cattle grazing near surface waters; buffer strips for spray drift; discharge consents from in-plant wastewater treatment works to the sewer); or (iii) through chemical substitution within products. This said, some risks have to be managed post-supply or discharge; such as the pathogenic risk that exists in drinking water distribution systems for example, which is managed through maintenance of a chlorine residual post-treatment.

Credit: istockphoto © oticki

The early screening and prioritisation of risks can assist in focusing the mind of the risk analyst on risk critical aspects for further study. The objective of risk screening is to seek to establish, early on:

- the environmental setting of the study;
- which hazards and adverse effects the risk assessment is principally concerned with;
- the key sources (of the hazard), pathways and receptor components of the risk;
- the existing, or potential connectivity between sources-pathways-receptors;
- the extent to which the risk is driven by the potency of the source, and/or the
- likelihood of exposure and/or the vulnerability, sensitivity and/or number of receptors;
- the key uncertainties or unknowns.

Hazard identification is a critical first step at whatever tier of analysis is being conducted (right hand embedded box; Figure 4.1). Registers can be used during hazard identification to formalise the process of recording what might go wrong if hazards are realised. Following an initial brainstorm involving a team of relevant experts, hazards are listed and assessed using the judgment of a knowledgeable

INITIATING EVENTS, EXPOSURES AND HARM. SOURCES, PATHWAYS AND RECEPTORS

A comprehensive review of 98 waterborne disease outbreaks (Hrudey & Hrudey, 2004) provides a list of six principles for providing safe drinking water:

(1) Pathogens (**source**) pose the greatest and most tangible risk to drinking water safety (**exposure of receptors**), making pathogen removal (**pathway**) and disinfection (**pathway**) the paramount concern.
(2) Robust, effective multiple barriers (**pathway**) to drinking water contamination are needed to suit the level of contamination challenge (**source**) facing the raw water source.
(3) Trouble is usually preceded by change – so change should be taken as warning to be on alert for trouble (**initiating events**).
(4) Operators must be capable and responsible.
(5) Drinking water professionals (providers, regulators and health officials) must be accountable to drinking water consumers (**receptor**).
(6) Ensuring safety is an exercise in risk management, requiring sensible decisions in the face of uncertainty.

individual or the consensus view of the team. This is the approach adopted in a HAZOP (unit 3).

Another qualitative approach to risk screening is source-pathway-receptor-harm analysis. Here, existing, or potential linkages between these discrete components of a risk are set out in tabular form with reference to a schematic or conceptual model, which summarises these relationships visually. This approach provides a powerful means of screening and, by reference to the **potency of the source**, the **availability of the pathway** and **sensitivity of the receptor**, of prioritising environmental risk (Figure 4.3).

Following an initial qualitative assessment, the significance of risks can be evaluated using pre-defined and agreed indicators, such as 'high', 'medium', or 'low', or by using agreed scales of relative or absolute probability and consequence. Beyond the value in characterising the significance of the risk, this also helps identify the type and degree of risk management options that are appropriate. The use of semi-quantitative indicators to 'score' probability and consequence can offer insight into whether the significance of the risk is driven by the probability of their occurrence or by their consequences if they do occur. This helps to identify different approaches to risk management, allows a comparison of residual risks with those prior to the adoption of risk management measures and can inform the focus of more detailed studies.

Care needs to be taken when using numerical scoring systems, however, so as to avoid giving a spurious impression of accuracy. It is often preferable to use a structured team approach for carrying out a risk ranking exercise than for the risk assessor to undertake this alone. This enables the views of different experts to be incorporated and provides opportunities to involve a range of stakeholders. It can instil greater credibility and encourage, though not guarantee, greater acceptance of the results.

4.3 RISK MANAGEMENT FRAMEWORKS IN THE WATER UTILITY SECTOR

Risk assessments for drinking water originated in the United States (US) in the early 1970's. Amendments to the US Safe Drinking Water Act in 1974 demanded better estimates of potential hazards for risk management purposes. This was followed by a series of studies from the National Research Council, including the 1983 'red book', which provided a set of formalised steps for assessing risks to human health from chemicals in drinking water:

- **problem formulation and hazard identification** – to describe the human health effects derived from any particular hazard (e.g. infection, carcinogenicity, etc.);
- **exposure assessment** – to determine the size and characteristics of the population exposed and the route, amount, and duration of exposure;
- **dose-response assessment** – to characterize the relationship between the dose exposure and the incidence of the health effects; and
- **risk characterization** – to integrate the information from exposure, dose-response, and health interventions in order to estimate the magnitude of the public health problem and to evaluate variability and uncertainty

This approach became the basis for other risk assessment frameworks, including the US Environmental Protection Agency's (USEPA's) ecological risk assessment framework and the International Life Sciences Institute microbial risk assessment framework. In both these, the risk of a pathogen or contaminant is analysed along a 'source-pathway-receptor' route, using models where appropriate. This was further developed by the US Presidential Commission on Risk Management that reported in 1997. Critically, it recognised that *"the output of risk assessment … often seems too focused on refining assumption-laden mathematical estimates of small risks […] rather than on the overall goal, risk reduction and improved health status"*. The Commission examined the role of risk assessment within a broader health and

The source of the hazard	Hazard – the substance or situation with a potential to cause harm	Receptor – the feature we value	Transport mechanism – how the source of the hazard reaches the receptor	Exposure pathway	Probability of exposure[1]	Consequences[2]	Risk[3]	Justification for this assessment of risk
Sewage pumping station	Routine odour releases from the pumping station impact	Public; local community; maintenance workers	Transport in the air and dispersion off-site	Inhalation	High	Moderate	High	Routine exposure; intermittent release, high impact on occurrence; maintenance works present regularly.
Overspill of sewage to ground, with percolation through subsoil to aquifer; downgradient flow to abstraction borehole	E.Coli and associated pathogens	Humans; those drinking groundwater downgradient	Percolation through soil; transport in aquifer	Ingestion of contaminated water from borehole; no treatment required	Negligible	Negligible	Low	Borehole at considerable distance; sand bed;
Pumping station overspill to Sea, via pipe; no barrier	Pathogens in raw sewage overspill	Marine life	Via pipeline and direct discharge; no treatment	Direct contact	Low	Moderate	Medium	
Pumping station overspill to Sea, via pipe; no barrier	Pathogens in raw sewage overspill	Human, recreational water users	Via pipeline and direct discharge; no treatment	Ingestion	Low	High	Medium	Intermittent discharge typically twice per year;
	Biochemical oxygen demand in raw sewage	Marine life	Via pipeline and direct discharge; no treatment	Direct contact	Low	High	Medium	

[1]Probability of exposure is defined as the likelihood of the receptors being exposed to the hazard. **High**: direct exposure likely with no/few barriers between hazard source and receptor; **Medium**: feasible exposure possible – barriers to exposure less controllable; **Low**: several barriers exist between hazards source and receptors, to mitigate against exposure; **Negligible**: effective, multiple barriers in place to mitigate against exposure.

[2]The consequences of a particular hazard being realised may be actual or potential harm to human health, incorporating spatial and temporal extents of potential harm and reversibility. Assumes child as most sensitive human receptor. **Severe**: there is sufficient evidence that short- or long-term exposure to chemical may result in serious damage to health (e.g. death, clear functional disturbance or morphological changes which are toxicologically significant). Latency of effect and irreversibility (during or following exposure) should be considered here; **Moderate**: there is sufficient evidence that exposure to chemical may result in health effects that are not severe in nature and are reversible once exposure ceases (e.g. irritant); **Mild**: Health effect not apparent though chemical exerts reversible physiological and/or pathological changes (e.g. biochemical, haematological changes or enzyme induction but no other apparent effect); **Negligible**: No evidence of adverse health effects and/or physiological and pathological effects following exposure to chemical.

[3]Qualitative evaluation of the significance of the risk determined by combining the probability of the consequences (i.e. probability of (a) the hazard occurring; (b) the receptor being exposed to the hazard and (c) harm resulting from that hazard) and the magnitude of the consequences.

Figure 4.3 Illustrative qualitative risk assessment. The environmental risk from sewage pumping station overflow.

Environmental risk assessments for chemicals in catchments are based on comparing exposure with an effect threshold (Predicted No Effect Concentration: *PNEC*). For so-called "down-the-drain" chemicals (pharmaceuticals and ingredients in household cleaning products), exposure assessments (i.e. calculations of Predicted Environmental Concentration: *PEC*) are based on a simple ratio of per capita consumption and per-capita domestic water use, adjusted for removal during sewage treatment and for dilution in the receiving environment using a dilution factor (e.g. 10). One problem with this is that it does not take into account spatial and temporal variability in the dilution, which changes with the relative magnitude of point-source loads and river discharge at the point of emission. Higher tier exposure models such as GREAT-ER (Geography-referenced Regional Exposure Assessment Tool for European Rivers) can predict the distribution of reach-specific concentrations, accounting for time varying emissions and in-stream dilution and degradation. Predictions can be compared with effect thresholds or integrated with distributions of effect end-points in a range of organisms to identify risk "hot spots" or predict overall risk.

environmental context, setting out a framework for risk-based decision-making and seeking to highlight that risk management is an iterative process responsive to new information that changes the need or nature of the assessment. It formally recognised the central role of involving stakeholders – individuals, groups or organizations, who can affect or who are perceived to be affected by the risk – early on in the assessment process at the problem definition or 'framing' stage.

The literature contains many risk management frameworks for risk-based decision-making. Their key features are:

- a problem formulation stage;
- stakeholder involvement;
- communication;
- quantitative risk assessment components;
- iteration and evaluation;
- informed decision-making; and
- flexibility.

An important milestone was publication of the Australian/New Zealand Standard 4360:1999 generic risk management standard. This made the case for effective risk management being an integrated part of everyday business activity and it led to design of the international standard ISO 31000 in risk management (Unit 6). Critically, it recommended that companies should identify:

- the roles and responsibilities of various parts of the organization in managing risk;
- relationships between the project and other parts of the organization; and
- the goals, objectives, strategies, scope and parameters for risk management.

4.4 RISK ASSESSMENT – CATCHMENT TO TAP

4.4.1 Assessing risks in catchments

Catchment (or watershed) management has gained widespread international support and is informed by the risk assessment of activities within catchment. In Europe, the 'DPSIR' approach to identifying key hazards within a catchment, by reference to the Drivers, Pressures, State, Impacts and policy Response (the risk management strategy) is adopted under the European Water Framework Directive. These terms are defined and exemplified in Figure 4.4.

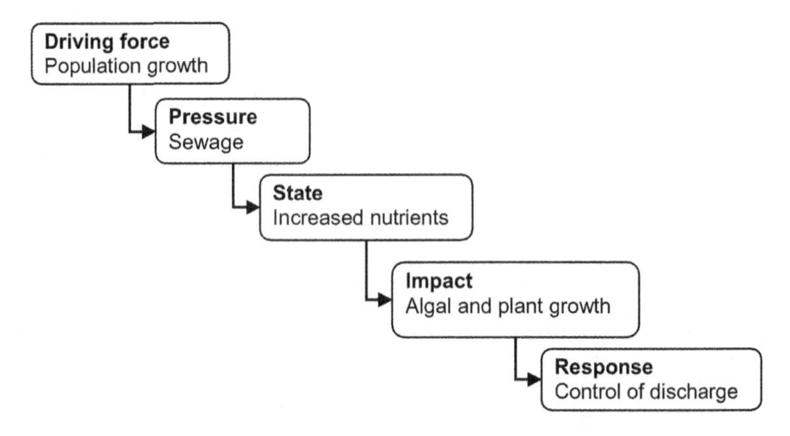

Figure 4.4 DPSIR concept illustrated for anthropogenic eutrophication.

- **driver** – an anthropogenic activity that may have an environmental effect (e.g. the intensification of agriculture; urban sprawl; population growth);
- **pressure** – the direct effect of the driver (an effect that causes a change in flow, or a change in the water chemistry;
- **state** – the condition of the water body resulting from both natural and anthropogenic factors (physical, chemical and biological characteristics);

- **impact** – the consequences that the pressure exerts on the state of the water body;
- **response** – the measures taken to improve the state of the water body (restricting abstraction, limiting point discharges, developing best practice guidance for agriculture).

Here, assessments of activities posing an actual or potential threat to the quality of water bodies in 'river basin districts' are used to prioritise a programme of multi-agency action plans targeted at raising the ecological status of the watershed within statutory timescales. Given the myriad of potential management issues for any catchment improvement programme, there is a need to prioritise risk management efforts in the watershed by concentrating on measures that will exert the greatest reduction in the likelihood of severe impacts (harm) being realised. The explicit consideration of probability has not conventionally been an inherent feature of DPSIR analysis because causality between the DPSIR components is assumed. There is a need to consider addressing this in order to improve the power of analysis.

Given the capabilities of modern geographical information systems (GIS), and their ability to map and analyse data that is spatially and temporarily variable, many catchment-level ranking methodologies incorporate overlay techniques, (combining the attributes of two or more data layers across geographic space), in the mapping of areas critical to catchment water quality. These spatial risk-mapping methodologies deploy attributes considered to play a significant role in, say, pollutant transport, (e.g. geology, rainfall, soil type, agricultural activities *etc.*), according to pre-defined formulae, (e.g. a weighted runoff-potential index). As with all risk assessment techniques, they may be generic or targeted towards specific hazards, (e.g. animal feeding operations), or pollutants, (e.g. through incorporating measures of their leaching potential).

Where a more detailed analysis is necessary, water quality and flow modelling is used to model the dispersal of pollutants and predict the resultant deterioration in water quality. Aside from the inherent value of fostering an increased understanding of catchment water quality issues, the benefits of model-based analysis stems from their ability to test risk management scenarios (through e.g. sensitivity and scenario analysis), thus enabling informed decisions on how best to manage the resource.

GEOGRAPHIC INFORMATION SYSTEMS

(GIS) play an increasing role in asset management. GIS systems have evolved into extensive inventories of asset condition records, enabling failure and maintenance data to be collated for capital improvement planning and maintenance works. GIS also support studies of distribution network corrosion, based on spatially variable corrosion risk factors as a causal factor in water mains deterioration. In relation to mains failure, GIS has been applied in probability studies based on contributing factors, e.g. pipe number density, pipe diameter, soil corrosivity, to correlate with past failure data. Finally, GIS have been embedded with hydraulic modelling capabilities to assess the potential of intrusion in distribution systems. These consider adverse pressure gradients and contaminant sources in identifying areas susceptible to intrusion.

4.4.2 Process risks – the multi-barrier concept

Raw water sources are exposed to a wide range of chemicals and pathogens that pose a potential health hazard to the consumer. Contamination may arise from many diverse sources, including pathogens, industrial waste, (organic and inorganic chemicals), by-products of agricultural practices, (e.g. fertilisers, insecticides, pesticides *etc.*), and leisure activities (petroleum products). Any chemicals present in the raw water supply will be present in the process water from the intake and subjected to the various stages of water processing.

The principal objective of water processing is to produce good, safe drinking water that has the trust of customers and is produced within acceptable drinking water quality parameters, including:

- coliforms
- faecal coliforms
- colour
- turbidity
- odour
- taste
- hydrogen ion (pH)
- nitrite
- aluminium
- iron
- manganese
- lead
- trihalomethanes
- total pesticides
- individual pesticides
- nitrate

This list is not comprehensive and other chemicals, emerging contaminants and pathogens may be present in the water supply that present a risk.

Some chemicals are also deliberately added during water treatment, such as aluminium sulphate, to promote flocculation and precipitation of solids and chlorine for disinfection. Risks may also arise as a result of processing. For instance, chlorination introduces by-products, such as trihalomethanes (THM) that are toxic at sufficiently high doses. Human error may also lead to the accidental introduction of chemicals directly into the distribution system as demonstrated by the Camelford incident in Cornwall where alum was added directly into the potable supply causing health problems and raising the fear of increased likelihood of Alzheimers disease in the local population.

The multiple barrier approach has been used to describe the effects of treatment processes on contaminants entering a water or wastewater treatment works. It relies on the concepts of treatment efficiency and Reliability; process Availability; and system Maintainability (RAM). For example, the risk of disease, infection or fatality, arising from the presence of pathogens, can be reduced by the use of filtration and disinfection processes, such as chlorination.

EXAMPLE 4.1 TREATMENT BYPASS

Bypass occurs when pathogens evade treatment. Remember from Unit 3 that bypass has a dramatic effect on exposure, such that with a 6-log destruction of pathogens for 99% of the time, just 1% bypass reduces the removal to 2-log. The integrity of unit processes for pathogen removal is critical because if bypass persists, increasing the number of unit processes prove futile.

The concentration of an undesirable microbial species is expected to fall as a result of water treatment. However the efficiency and effectiveness of chlorination, for example, will not be the same for all microbial species and water qualities. To a first approximation, the concentration of microbial species present in water following, say, disinfection may be represented using the following expression:-

$$C_2 = (1 - E \cdot A)\, C_1 \qquad (4.1)$$

where C_2 is the effluent concentration of species following disinfection; C_1, the influent concentration of species into the system; E, the efficiency of the disinfection system; and A its steady state availability of the unit process.

Equation 4.1 can be used to estimate the impact of processing stages that reduce the concentration of contaminants. For a two stage process involving filtration (a) and disinfection (b), with an influent concentration C_{infl} and an effluent concentration C_{effl} ,the effluent concentration is predicted to be:

$$C_{effl} = C_{infl}\,\{(1 - E_F)\,(1 - E_D A_D)\} \qquad (4.2)$$

If the efficiency of disinfection $E_D = 0.99$ and of filtration is say 0.90 and the availability (mean time in the operating state) of the disinfection system is 0.9 then the effluent concentration is $C_{effl} = 1.1 \times 10^{-2}\, C_{iinfl}$.

Equation 4.2 indicates that the greater the number, efficiency and availability of processing stages, the lower the effluent concentration. The effluent concentration can be compared either with the maximum contaminant level set by regulators or with the toxic dose response curve to estimate the risk of infection. The concentration of species at the point of use will depend on how the water distribution network affects the stability of chemicals and the growth of microbial species. At the point of use the actual concentration may be greater or less than that leaving the plant.

EXTENDED MULTIPLE BARRIER APPROACH:

- source protection
- multistage treatment
- secure distribution system
- intelligent monitoring to inform risk management
- thorough and effective response capabilities

The possibility of microorganism survival, re-growth and additional load within the distribution system is managed by overdosing with chlorine that in turn yields a chlorine residual. However, because the by-products of chlorination are also toxic, there has to be an upper limit as well as a lower limit on the amount of chlorine added to water, generating a risk trade off. Risk analysis can be used to establish allowable concentration limits for microorganisms and residual chlorine, once risk tolerance levels have been established.

An extended conceptualisation of the multiple barrier approach applies to the full water supply chain. Here (see box) catchment to tap controls and procedures are considered.

4.4.3 Network risks, vulnerability and Markov models

Infrastructure networks pose a particular type of challenge for risk analysis because they are conventionally buried, difficult to survey and manage and the costs of maintenance and/or replacement are high. Increasingly, we are concerned with networked and interconnected infrastructures with cascading failures and the potential for systemic collapse, say from power, or IT outage, for example.

A water distribution system is an interconnected network of sources, pipes, and hydraulic control elements (e.g. pumps, valves, regulators, tanks) delivering water to consumers in prescribed quantities and at desired pressures. System behaviour, which is governed by hydraulics, supply, demand and system layout, can be described mathematically and has been discussed in length within Unit 3. This forms the basis of water supply and distribution modelling (network analysis), a discipline practised in the water industry for many years, particularly to inform the development of operational RAM strategies.

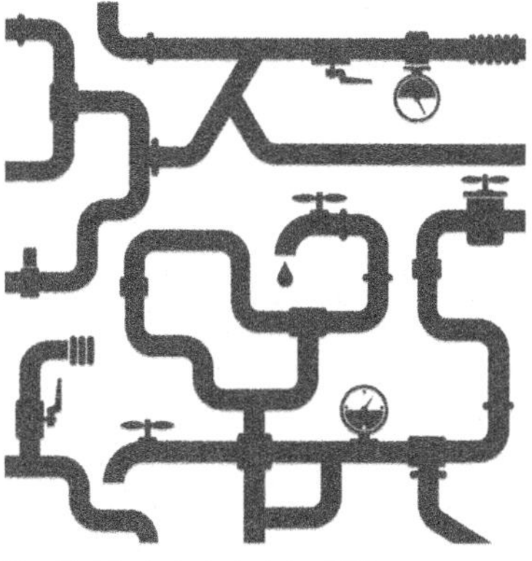

Credit: istockphoto © John1179

Water utilities routinely apply network analysis to assess their 'security of supply', defined as the probability of being able to meet consumer demands (*i.e.* network reliability). Best practice utilities extend this analysis beyond routine operating conditions to evaluations of network performance under various supply-demand scenarios, reflecting the inherent uncertainty of the supply-demand balance.

Typical among these models is the resource modelling package 'WRAPsim' that has been applied in the UK. In one example of its application, the model is configured to contain over 1200 components including all river and reservoir sources, boreholes, water treatment works, pipelines and demand centres. Through simulation of a utility's sources over a given time period, the model provides a decision-maker with an accurate assessment of the behaviour of each source, its ability to meet demand, and the frequency of restrictions that would need to be imposed. Further insights are gained through the application of scenario analysis, wherein the supply-demand balance for each zone under variable scenarios (e.g. average year, dry year, peak week, *etc.*) allows an assessment of the security of supply over a range of timescales and operating conditions. The capacity of these tools to predict future supply conditions, to optimise the allocation of water resources and to inform the rebalancing of stocks may significantly increase the yield and reliability of a water utility's supply system.

Asset management tools are used widely for distribution networks and underground assets. For example, many sewer deterioration models are adaptations of event trees (Unit 3) that consider the conditional probability and consequences of a sewer collapse in successive years' service, presuming a progressive reduction in condition over time. So-called Markov models are used to quantify these processes and have much in common with event trees. A Markov chain describes the states of a system over successive time intervals, (say, at yearly intervals; Figure 4.4). It is assumed that, within the time interval, the system may have either changed or stayed in the same state. Markov changes of state are referred to as 'transitions' and the conditional probability distribution of the system's state in the future depends only on its current state, and not on its state in the past - the output of a Markov analysis is a set of possible states, (conditions of a sewer), and an attending set of probabilities (Figure 4.4).

In order to complete a Markov model of sewer collapse, one needs to know the transition probabilities – the probability of moving from one condition grade to another (of lesser condition) in any one year – and the probability of catastrophic collapse in any one year from any particular condition grade (Figure 4.5).

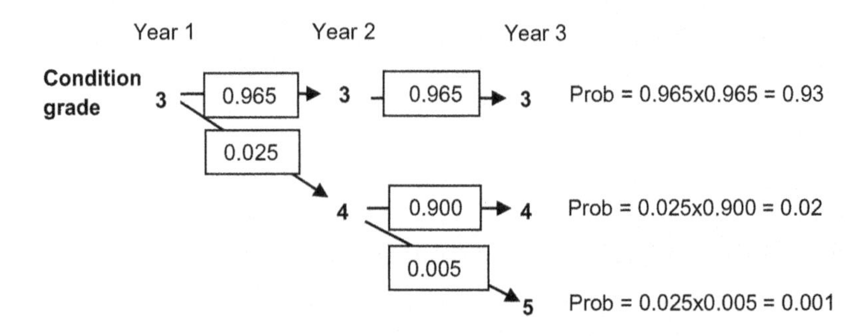

Figure 4.5 Partial Markov sewer deterioration model. In successive years 1 to 3, the sewer can follow a number of future condition sequences, contingent on its maintenance, for example. In this example, the sewer may be maintained at condition grade 3 (condition grade 1 is high; 5 is low and deteriorated) or in Year 2 deteriorate to condition 4. Year 3 may see further deterioration to condition 5. Individual transition probabilities between successive years and their combination allow a calculation of the probability of a fixed length of sewer deteriorating from one condition to another over a specified time period.

Network analysis can: (a) allow utilities to assess their susceptibility to various supply-demand scenarios, (e.g. drought or increases in demand); (b) aid decision-makers in determining 'optimal' supply strategies and policies; (c) assist in the design phase of distribution networks; and (d) inform the need for capital expenditure.

Operational disruptions are the inevitable result of large-scale disasters, (e.g. flooding, drought, earthquakes, and terrorism). To minimise the risks posed by uncontrollable events, utilities seek to eliminate or reduce their potential consequences through contingency and emergency planning. The role of formal risk analysis in emergency planning, once restricted to drought management, is now widely adopted to address security risks. Methodologies for appraising system vulnerabilities and for developing measures to reduce the risks and mitigate the consequences of terrorist or other criminal attacks include:

(1) determining how well the system detects a problem, which involves surveying all security and monitoring features, (e.g. how quickly could it detect an undesired chemical being introduced to the supply);
(2) measuring delay capabilities in order to determine how well a system can stop undesired events, (e.g. security in place, length of storage time); and
(3) measuring the capacity of private guard forces and local, state and federal authorities to respond to an event.

4.4.4 Public health risk

Here we are concerned with the risk posed by specific contaminants to human health and the related risk of exceeding regulatory standards. The principal sources of water quality problems within the house are elevated concentrations of metals (lead, copper and zinc) from pipe corrosion, taste and odour problems and pathogens. Corrosion can take various forms, ranging from gross metal loss to corrosion-assisted cracking and fatigue when mechanical stresses are present. The equipment most likely to be affected by corrosion will be those components manufactured from, or containing, steel or cast iron parts. With the replacement of iron, lead and copper pipework with plastic pipes, chronic health risks from corrosion have become less of an issue for in house systems.

Pathogen load should be reduced normally through maintenance of a chlorine residual within distribution systems. The multiple barrier approach to water treatment has been the central tenet of modern water treatment systems and relies on the use of 'in-series' treatment processes to remove hazardous agents from the public supply. Failure, or the inadequacy of the treatment and distribution process can result in a derogation in water quality, (microbiological or chemical), with potential impacts on public health. The underlying causes of failure may include source contamination, human error, mechanical failure or network intrusion. Within distribution networks, there has been a concern that occasional negative pressures in distribution pipes may lead to microbial intrusion and a subsequent pathogenic load introduced into the treated water being distributed. The consequences of failure are often immediate – there is little time, if any, to reduce exposure because of the lag time in securing meaningful monitoring data and impacts can often affect a large number of people simultaneously.

Beyond these impacts on public health through the direct ingestion of contaminated drinking water, financial and consumer confidence impacts invariably ensue. The financial costs to the community of the fatal Walkerton outbreak, for example, were in excess of Cdn$65m, with one time costs to Ontario estimated at more than Cdn$100m. Compounding this, the loss of consumer confidence following disease outbreaks can be enormous. Even when there is no legislation covering aspects there can be claims of negligence against operating companies. Litigation for civil damages has been prominent features following both the Walkerton outbreak (settled out of court) and the Sydney Water crisis (largely dismissed, costs still incurred).

4.5 HUMAN FACTORS

We mentioned at the beginning of this unit the importance of behaviour as a means to managing those risks that extend beyond strict engineering processes. In fact, all technological systems harbour engineering, human and organisational features subject to potential failure with the onward consequences of harm to human health and/or the environment. The discussion of water quality incidents raises the issue of the role of human factors in risk assessment and management. As in most business sectors, the majority of accidents and incidents in the water sector have a strong human dimension (Figure 4.6).

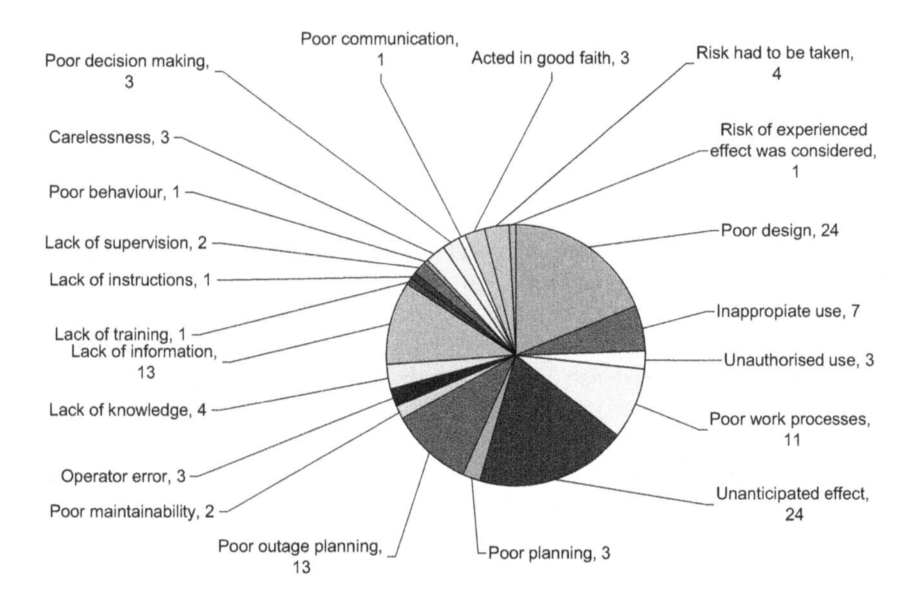

Figure 4.6 Human factor-related incident causes for incidents between 2004 and 2006 (after Bradshaw *et al.*, 2008).

Any quantification of risk and system reliability rests heavily on the availability of data for hardware, software and the human components of a system. The lack of accurate human reliability data is a serious limitation in all risk assessments and major source of uncertainty that is frequently ignored. Below, a number of incidents that have occurred in the water utility sector are described and related to current theories of accidents causation. Finally, one basic method of estimating human reliability and the likelihood of human error is described, as a means of exploring the complexity of human factors in risk analysis. Readers should consult Hrudey and Hrudey (2004) for in depth analyses of disease outbreaks. Consider the following three incidents, each of which has notable engineering, human and organisational features and broader societal consequences:

4.5.1 Overdose of aluminium sulphate

A tanker driver unloaded 20 tonnes of concentrated aluminium sulphate solution into the wrong tank at an unmanned water treatment works. The receiving tank was the treated water reservoir that supplied 20,000 people many of whom claimed they were subjected to either short or long term illness. There have been a number of court cases and claims from people who drank the water. Since the accident, the organisation has been privatised, but with considerable difficulty. The accident diverted substantive management time to restoring customer confidence and taking a number of measures that should have been better planned or need not have been as extensive.

Factors leading to the accident:

- it was the tanker drivers first visit to the site;
- the driver was given a key to the site by the regular driver;
- the treatment works operative had decided to take some time off work;
- the works was being modified and accesses to tanks were not clear;
- tanker deliveries were carried out by lifting a manhole and allowing the tanker delivery hose to discharge into it;
- there was no recording procedure for the delivery, so that when customer complaints were received there was no immediate link with the delivery;
- as the level of the alum tank was low it was assumed the delivery had not taken place.

Consequences of the accident:

- 20,000 people received drinking water with high levels of aluminium for three days;
- the acidic water leached out metals from the pipes carrying it, so that the levels of copper and zinc in the drinking water were also very high;
- customers complained of:
  - extremely unpleasant taste
  - blue colour
  - curdled tea and coffee
  - blue – green scum forming when soap was used
  - staining of basins and baths
  - staining of finger nails and hair
- a number of customers made major claims against the water utility in court;
- there were extensive fish deaths when the water was flushed from the mains into the water courses;
- the incident attracted high media coverage by television and press;
- the company took many actions to ensure the incident did not reoccur including:
  - many changes to its operational staff
  - risk assessments at water treatment works
  - many changes to operational procedures
  - adopted new emergency procedures
- it was extremely difficult for the government to privatise the company 15 months after the accident.

Procedures that should have been in place:

- site keys are only issued to specified employees (and never to drivers);
- deliveries only take place with the plant attendant present;
- tanker hoses should have a coupling point which has fittings agreed for only that chemical;
- coupling points should be clearly marked;

- tanker contents should be sampled before discharge;
- delivery notes should be handed to the plant attendant;
- the treated water outlet to the works should have sufficient analysis to identify deviations from the target parameters.

4.5.2 Loss of water supply to a city

A supply contained several pipelines of 1.0 m to 1.2 m external diameter that had been laid as the system was developed. Generally the pressures were low and there were a number of valves cross-connecting these pipelines and connecting it to other pipelines. Initially a small amount of flooding was observed in a playing field. The effected length was isolated and arrangements made to commence repairs to the pipeline. Subsequently a much larger burst occurred resulting in extensive flooding and a supply outage with 140,000 people having no water supplies for up to 8 days.

Factors leading to the incident:

- the wrong valve was closed during the initial isolation leading to the introduction of water at much higher pressures into the second pipeline, which as a result, burst;
- some valves failed to operate;
- there were inadequate layout plans of the system;
- valves were not marked to identify them;
- there had been a staff reduction exercise and the employees with the local knowledge were no longer employed by the organisation;
- for financial reasons maintenance had been reduced to a minimum, which contributed to the lack of operation of some valves, and a lack of local knowledge and drawings;
- a Major Incident Procedure Plan was not in place and therefore there was insufficient management capability available to respond to the incident, which developed after the second burst.

Consequences of the accident:

- people lost their water supply, many for 8 days, this increased the risk of diseases and illness;
- two areas suffered flooding;
- the repair costs for the two bursts, plus the cost of restoring the water supplies was high;
- outside organisations, including the Army had to be brought in to the City to provide essential drinking water supplies by other methods;
- factories and hospitals were put at risk during the emergency;
- an extensive operation to locate, mark, plot valves onto system maps and carry out maintenance on them was started;
- the organisation was slow in meeting claims for damage;
- there was considerable television and press coverage.

Procedures that should have been in place:

- adequate training and the development of the operator's knowledge and expertise of the system;
- plans of the pipelines with the location of valves, PRVs etc. and operating constraints such as maximum pressures;
- valve location signs;
- valve maintenance programme;
- major Incident Management Plan;
- communication systems to ensure co-operation within geographical divisions of the organisation;
- procedures with the insurance company to allow prompt interim payment of claims for damage due to flooding;
- need to identify customers with special needs, including hospitals;
- need to manage the television and press.

4.5.3 Overdose of lime at a water treatment works

An electrical fault in a lime dosing plant, combined with a faulty pH monitor resulted in 15,000 local residents plus 15,000 tourists receiving water supplies with a pH of up to 12. The organisation only detected the problem following complaints. It took over 48 hours to clear the system during which time customers were advised not to touch, drink or cook with the water.

Factors leading to the incident:

- an alarm on the plant during the night when it was unmanned resulted in a visit by an operator. The operator found the rapid gravity filters were retaining a lot of lime;
- the operator carried out back-washing of the filters and because there was no other alarms he did not carry out any checks on the plant;
- the operator made a routine visit at 8 am, and found extensive deposits of lime on the filters. As there were no alarms, nobody else was informed and no investigations carried out to ascertain why the failure had occurred. (The attendant saw his role as making sure the water leaving the works was above minimum standards, and if it was below that he should take remedial action as soon as possible. He would only inform his supervisor if he was unable to correct any problem.)
- the works diary indicated that levels of pH above 9.5 (the legal limit) had occurred several times and the operator had taken steps to reduce it, but not reported it to his manager;
- there was no works operating manual or on-site emergency response plan which may have reduced the impact of the incident.

Consequences of the accident:

- 30,000 people were unable to drink or touch the tap water for up to 48 hours during hot weather;
- a massive operation was launched to provide alternative drinking and washing water;
- the time to flush out the distribution system was governed by the lack of water in the local service reservoirs. A major tankering operation was launched to provide additional flushing water
- as the organisation had a major incident plan in place the incident was well managed, with plenty of information to the press and direct to the customers. There were very few complaints about the response to the incident;
- additional monitors and alarms were installed at this and other works;
- the control system on the lime batching plant was modified to reduce the risk of incorrect batching due to a failure;
- temporary and subsequently permanent works operating manuals were produced which included on-site emergency response plans.

Procedures that should have been in place:

- a procedure by the operator to check whether there is a second failure (in this case the pH monitors) when an initial failure has been found;
- a communication system to ensure any problems are reported so that they can either be linked with customer complaints or additional checks and advice sought if appropriate;
- the allowable range of all parameters should be clearly set out either in the works operating manual or on a board in the control room. All deviations from these set levels should be communicated to a designated senior member of staff.
- operators should be trained on the acceptable limits and the consequences of exceeding them;
- monitors should issue an alarm when they fail (i.e. fail-safe);
- works diaries should be monitored frequently by supervisory staff;
- management should have a better understanding of the ways in which the works can fail due to mechanical failure and human factors.

4.5.4 Assessing human reliability

Real systems combine people with technology and both technical hardware and human reliability have to be considered. The definition of human reliability and equipment reliability are similar, at least in basic concept, and this underlines one approach used in assessing systems; namely that the human is treated as a kind of component within the system who performs a set of discrete tasks, each with a certain reliability. One definition of human reliability then is: *the probability that a task will be successfully completed by personnel at any required stage in a system's operation within a required minimum time (assuming a time requirement exists).* Mathematically this can be expressed as:

$$HR = n_s/N_0 \qquad (4.3)$$

where n_s = The no. of successful completions of a given task
and N_0 = the number of attempts at a given task

The probability of human error (HEP) is given by:-

$$HEP = 1 - HR = n_f/N_0 \qquad (4.4)$$

Where: n_f = the number of failures

The error rate (or error frequency), l, is given by:-

$$l = n_f/\Delta t\, N_0 \qquad (4.5)$$

Credit: istockphoto © Terry J Alcorn

The human error rate is therefore analogous to failure rate of a hardware component (Unit 3). However, there are two significant differences between equipment and human reliability, namely:

- a human is often capable of error correction after an error has been made; and
- under stress, human reliability may increase or fall to a very low value depending on the level of stress.

Thus, the reliability of the human varies widely and is difficult to predict. It is possible, however, to introduce human error as a discrete event into fault tree or event tree analysis as part of a system study. However, this presupposes that a value for the probability of failure of the human action can be determined. Empirical estimation techniques are available to support this and are briefly introduced below.

4.5.5 How accidents happen

Before considering the quantification of human reliability in systems, it is important to understand how human behaviour influences accidents. The study of accidents provides important insights into the development of realistic human reliability models. The key stages of an accident are usually (i) an initiating event; (ii) loss of safety barriers/defences; (iii) deterioration of conditions and event escalation; followed by (iv) failure to evacuate or escape.

Firstly, the initiating event: Many accidents are triggered either by human error or the failure of a piece of equipment. The trigger event may be quite small, it may be deliberate or accidental and is often associated with routine, relatively insignificant activities such as maintenance work. The underlying causes of such events, however, are complex with a range of interacting human, organisational and hardware factors. These causes often lie deep within organisations.

Next, the loss of defences: The ability to control an event at an early stage in its propagation depends critically on prompt human reactions together with the availability and integrity of emergency response equipment and safety control systems. Such systems must be robust and capable of withstanding the 'loads' imposed by initiating events, as experience has shown that the initiating event may prevent some of the emergency control systems from functioning, resulting in a reduced capability to control the incident and a more rapid escalation of events.

Deterioration of conditions: This is the point at which an incident usually escalates from a minor to a major accident. In the case of fires and explosions the rate of escalation will depend on the scale of the initiating event and on the inventory of materials (e.g. flammability) as well as on the design and construction of the plant and surrounding buildings. Incidents that escalate rapidly increase the potential losses of assets and life – escape routes may become blocked or difficult to find and evacuation/escape of personnel may become more difficult.

Failure to escape or evacuate: The greatest impact on potential loss of life, the harm, arises for fires when evacuation or escape of personnel from a life-threatening situation is impaired. This usually makes the difference between accidents with fatalities and those with no loss of life. Evacuation and escape can be greatly facilitated by design e.g. provision of additional escape routes in buildings, protection of escape routes, temporary refuge areas, and transport systems to provide rapid evacuation etc. Evacuation drills are important to increase the rate of escape and evacuation from danger.

Wu *et al.* (2009) characterised the stages of a typical drinking water quality incident (Figure 4.7), the central significance of which is the onset of contamination prior to the sensing of abnormality or the issue of warnings. This reality is the rationale for a philosophy of preventative risk management as far as public health risks in the sector is concerned.

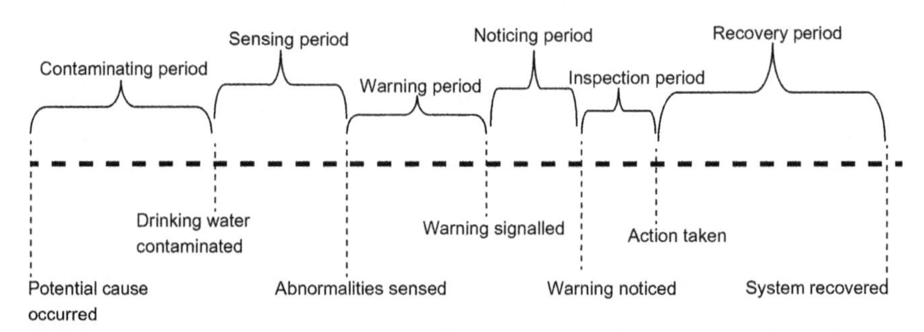

Figure 4.7 A typical gestation for a drinking water incident (Wu *et al.*, 2009).

4.5.6 Human interactions and classes of unsafe acts

Humans need to interact with complex systems of equipment, carrying out tasks that cannot be carried out by machines and instruments. Often these tasks imply some element of control. Human error involves a deviation from a presumed optimum action, sequence of actions or strategy. In many ways, human error is an inevitable consequence of our innate adaptability as humans. 'Adaptability' gives an individual the flexibility to work using a wide range of strategies to cope with changing circumstances. However, this ability can work against us in situations where a strictly non-redundant (discrete) sequence of operations has to be adhered to in order to avoid disastrous consequences. Psychologists argue that errors have causes and these can be intentional or unintentional. Most often errors are assumed unintentional – but some consideration has to be given to deliberate acts of error *i.e.* violations, circumvention and terrorism.

Field theories emphasise the necessity to consider the way in which individuals form a mental picture of a situation. The error potential is not realised until a predisposing condition creates a mismatch between the task and the ability to carry out the task. Too great a difference between the 'mental model', that is the understanding of the 'situation', and physical reality will bring about errors. This can result in a disaster when there is an inconsistency between 'the perception of what is occurring' and reality, if the system is in a critical state.

Accidents occur as a result of unsafe acts that breach system defences or occur in the absence of system defences. In a well-protected system, accident defence strategies such as awareness, recovery, protection and escape exist to prevent unsafe acts from becoming incidents and accidents. However, all system defences have some degree of vulnerability and no system is 100% reliable. The greater the degree of vulnerability (in the extreme there will be no system defence) the greater the likelihood that an unsafe act will lead to an accident.

For unsafe acts, psychologists argue these do not just happen, but are caused. The characteristics of the task involved and the environment play an important role in shaping human behaviour, and hence the probability of an unsafe act. Unsafe acts are regarded as human errors. There has been a great deal of research categorising human error introducing such terms as error of omission and error of commission (forgetting/inserting task steps that should or should not be present).

One current view is that unsafe acts are either; unintentional errors *i.e.* slips and lapses or intentional errors *i.e.* mistakes or violations. Slips and lapses are unsafe acts where the action/inaction carried out was not what was actually intended (sins of omission and commission belong here together with lapses of memory). Mistakes, on the other hand, are human failures where the intentions are wrong but are purposefully executed, for example, misdiagnosis of a fault in a component leading to a repair which is irrelevant to the failure mode. The belief is that the action carried out is correct when in fact it is incorrect. These types of error are often very difficult to discover. Violations are of two types; namely sabotage and circumvention. The former are actions performed with malicious intent whereas the latter involves no malice. Circumvention of procedure is usually carried out in order to overcome an organisational barrier. Often there is some direct benefit to be made by performing the circumvention like convenience, saving time, effort or money. Circumvention is the result of conflicting goals, is the cause of many safety problems in industry and is linked to the culture of the organisation.

CAN WATER UTILITIES BECOME HIGH RELIABILITY ORGANISATIONS (HROS)?

Bradshaw *et al.* 2011 completed an in-company analysis of water quality events 1997–2006.

- Each incident required organisational capacity to identify the incident, reduce its impact on customers and processes, and procedures to re-instate normal operations. The utility valued clear management objectives and ensured they were well understood within the organisation and by its partners. Staff were in the right roles with fit-for-purpose skills sets to work effectively.
- The utility was well positioned to manage incidents and many HRO principles were identifiable as management practice under 'trying conditions'. The observance of HRO principles contributed to the resilience of the organisation in maintaining a safe and reliable drinking water supply. The utility evolved its incident management structure to handle customer impacts effectively and expediently.
- A large proportion of incidents were associated with asset failures which defined the technical reliability of the water supply system. The utility invested heavily in control systems and adopted a risk-based view in resource allocation. In this sense, the water utility significantly differs from HRO theory.
- The frequency of incidents in the water sector indicates that utilities should be more forward thinking to improve the resilience of drinking water systems to human and organisational error. This would present a structured opportunity to learn from previous incidents to enhance asset risk and drive a preventative risk management culture.

4.5.7 Violation producing conditions (VPC)

Below is a list of factors that are widely accepted as causing violations in organisations. Many of these are linked to the organisational attitudes, group norms and management culture. These are sociological issues:

- poor safety culture;
- worker management conflict;
- poor morale;
- poor supervision and checking;
- inappropriate work group norms;
- misperception of risks;
- perceived managerial indifference;
- no pride in work;
- attitude problems:
  - "it won't happen to us"
  - "who gives a damn anyway";
- perceived license to bend the rules;
- 'can do' culture;
- excessive zeal.

4.5.8 Error producing conditions (EPC)

Researchers have also developed extensive databases of factors that promote unsafe acts and accidents, the three most important being; unfamiliarity, time stress, and noisy signals (uncertainty). These can be directly related to task and system factors such as task complexity, resource limitations and work rate. But underlying all the EPCs and VPCs are the issues of good safety and risk management.

- unfamiliarity;
- time shortage;
- noisy or confused signals/communications;
- poor man machine interface
- designer user mismatch (wrong mental model);

- irreversibility (unforgiving system)
- information overload;
- technique unlearning (of new procedure;)
- knowledge transfer;
- misperception of risk;
- poor feedback;
- inexperience;
- poor instructions;
- inadequate checking;
- substance abuse;
- educational mismatch;
- macho culture;
- physical capabilities exceeded;
- hostile environment;
- low morale;
- monotony and boredom;
- disturbed sleep patterns;
- externally paced tasks.

4.5.9 Latent failures and flawed management decisions

The conditions that produce errors and violations are generated by latent failures. Reason (1997) refers to these as 'system pathogens'. System pathogens exist because of the way the business is organised. Classic examples include lack of training and education, incompatible goals and poor communications. Others include:

- inadequate training;
- communications failure;
- incompatible goals, i.e. profit and production before safety;
- inadequate design;
- poor procedures;
- poor maintenance;
- hardware defects;
- poor housekeeping;
- management/organizational failures

4.6 SUMMARY AND SELF-ASSESSMENT QUESTIONS

This chapter has been concerned with broadening our discussion of risk assessment and management beyond the engineering reliability of unit treatment processes. The provision of safe drinking water is also affected by the management of catchment risk, network reliability and human factors within organisations. We return the organisational aspects of risk management in more detail in unit 6 during a discussion of risk governance. For now, however, you should:

- recognise that risk assessment extends beyond quantitative reliability analysis;
- understand that open environmental systems are more challenging to assess given the inherent uncertainties involved;
- have started to develop an understanding of human factors and their role in generating the active and latent conditions that act as precursors for undesirable incidents.

SAQ 4.1 From source, through treatment, on to distribution and consumer supply, describe individual barriers that contribute to the management of pathogenic risk in drinking water.

SAQ 4.2 Undertake a DPSIR analysis for the impact of intensive agriculture on nitrate loadings to groundwaters.

SAQ 4.3 What shortcomings does the mathematical modelling of human error embody?

SAQ 4.4 Compare and contrast the principles of management control for hazards in (a) catchments; (b) distribution systems.

4.7 FURTHER READING

American Water Works Association Research Foundation (2007). Risk Analysis Strategies for Credible and Defensible Utility Decisions. AwwaRF, Awwa and IWA Publishing, Denver, CO, 93pp.

Calow P. (1998). Handbook of Environmental Risk Assessment and Management. Oxford, UK, Blackwell Science Publications.

Department for Environment, Food and Rural Affairs and Cranfield University (2011). Guidelines for Environmental Risk Assessment and Management, Green Leaves III. Á Gormley, S. Pollard, S. Rocks and E. Black, Defra Report PB13670, London, UK, 81pp.

DETR, Environment Agency and IEH (2000). Guidelines for Environmental Risk Assessment and Management, Revised Departmental Guidance. The Stationery Office, London.

Gray N. F. (1999). Water Technology. An introduction for environmental scientists and engineers, Butterworth-Heinemann, London, 548pp.

Hrudey S. E. and Hrudey E. J. (2004). Safe Drinking Water. Lessons from recent outbreaks in affluent nations, IWA Publishing, London, 514pp.

Huysmans M., Madarász T. and Dassargues A. (2006). Risk assessment of groundwater pollution using sensitivity analysis and a worst-case scenario analysis. *Environmental Geology*, **50**, 180–193.

Kletz T. (2005). Learning from Accidents, 3rd edn, Elsevier, Oxford.

Koormann F., Rominger J., Schowanek D., Wagner J. -O., Schroder R., Wind T., Silvani M. and Whelan M. J. (2006). Modelling the fate of down-the-drain chemicals in rivers: An improved software for GREAT-ER. *Environmental Modelling and Software*, **21**, 925–936.

Linkov I., Satterstrom F. K., Kiker G., Batchelor C., Bridges T. and Ferguson E. (2006). From comparative risk assessment to multi-criteria decision analysis and adaptive management: Recent developments and applications. *Environment International*, **32**, 1072–1093.

MacGillivray B. H., Hamilton P. D., Strutt J. E. and Pollard S. J. T. (2006). Risk analysis strategies in the water utility sector: An inventory of applications for better and more credible decision making. *Crit. Rev. Environ. Sci. Technol.*, **36**, 85–139.

Macleod C. J. A., Scholefield D. and Haygarth P. M. (2007). Integration for sustainable catchment management. *Science of the Total Environment*, **373**, 591–602.

OECD (2011). The OECD Environmental Risk Assessment Toolkit: Steps in Environmental Risk Assessment and Available OECD Products. The Organisation for Economic Co-operation and Development (OECD), Paris.

O'Hagan A., Caitlin E., Buck C. E., Daneshkhah A., Eiser J. R., Garthwaite P. H., Jenkinson D. J., Oakley J. E. and Rakow T. (2006). Eliciting Distributions – Uncertainty and Imprecision, in Uncertain Judgements: Eliciting Experts' Probabilities. John Wiley and Sons, Chichester.

Phillips K. P., Foster W. G. Leiss W., Sahni V., Karyakina N., Turner M. C., Kacew S. and Krewski D. (2008). Assessing and managing risks arising from exposure to endocrine-active chemicals. *Journal of Toxicology and Environmental Health – Part B: Critical Reviews*, **11**, 351–372.

Pollard S. J. T., Yearsley R., Reynard N., Meadowcroft I. C., Duarte-Davidson R. and Duerden S. (2002). Current directions in the practice of environmental risk assessment in the United Kingdom. *Environmental Science & Technology*, **36**, 530–538.

Pollard S. J. T. (2006). Risk management for the environmental practitioner, IEMA Practitioner No. 7, Best practice series, Institute of Environmental Management & Assessment, Lincoln, UK, 75pp.

Pollard S. J. T., Smith R., Longhurst P. J., Eduljee G. and Hall D. (2006). Recent developments in the application of risk analysis to waste technologies. *Environment International*, **32**, 1010–1020.

Price O. R., Williams R. J. van Egmond R., Wilkinson M. J. and Whelan M. J. (2010). Predicting accurate and ecologically relevant regional scale concentrations of triclosan in rivers for use in higher-tier aquatic risk assessments. *Environment International*, **36**, 521–526.

Reason J. (1997). Managing the Risks of Organisational Accidents. Ashgate Publ., Brookfield, Vermont.

Tang Y., Wu S., Miao X., Pollard S. J. T. and Hrudey S. E. (2013). Resilience to evolving drinking water contaminantion risks: A human error prevention perspective. *J. Cleaner Production*, **57**, 228–237.

Wu S., Hrudey S. E., French S., Bedford T., Soane E. and Pollard S. (2009). A role for human reliability analysis in preventing drinking water incidents and securing safe drinking water. *Water Research*, **43**, 3227–3238.

Unit 5

Regulating water utility risks

MODERN REGULATION

Modern regulation seeks to target regulatory resources on activities posing the greatest risk to public health and the environment. Regulators seek to understand the preventative risk management capabilities of utility companies and their corporate commitment to managing risk.

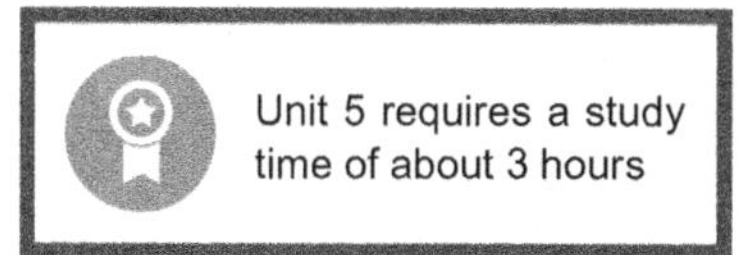

5.1 INTRODUCTION

Regulation has been widely used to reduce and manage public risk, and to protect the ecosystems on which we depend from harm arising from human activities. Government action may be required at a local, national and/or international level in order to secure this. While regulation is intended to deliver net social benefits, it also imposes costs on businesses and individuals, including the costs of administration, costs arising from changes to business practices such as installing new, cleaner technologies, and potentially indirect costs arising from constraints on competitiveness or innovation imposed by regulation. Regulation therefore necessitates a careful balance between incentivising protection and the costs of compliance. Modern, 'smarter' regulation seeks to deploy light touch measures on high performing businesses, whilst retaining the threat of firm enforcement on those that display flagrant disregard for the legislation.

Regulation has historically been viewed as a backstop to good operational practice within business sectors. It is there to ensure risks are managed in the public interest and that industrial and other activities face stakeholder scrutiny and compliance before, during and at the end of their operation. Modern regulation, rather than being adversarial, is better viewed as a partnership between operators, governments and their agencies, facilitated by the regulator. Together, and with their different preferences, priorities and interests, utilities and their regulators work closely together to ensure:

- the economic viability of operations;
- compliance with the statute and regulations;
- the capital maintenance of assets;
- the affordability of services for water and wastewater treatment; and most importantly
- the protection of public and environmental health.

REGULATION AND BEHAVIOURAL CHANGE

The extent to which regulated parties such as water utilities change behaviour in response to regulation depends on their understanding of the implications of regulation and an ability to manage risks and business opportunities. 'Organisational maturity' models for assessing this capacity have been developed by risk management institutions (AIRMIC/ALARM/IRM), and for specific sectors, such as the offshore industry and the water industry. Capacity building instruments like this enable organisations to change behaviour. Other instruments may increase capacity, for example, by providing clear standards that serve to educate the regulated in better environmental management, or through advice and guidance provided by regulators.

MARKET FAILURE AND REGULATION

One cause of economic inefficiency is when the financial returns a firm receives for performing an operation differ from the returns to society. The term 'market failure' describes a situation where, for one reason or other, the market does not achieve economic efficiency. Regulation is invoked to correct this; for example, for the externalities associated with environmental pollution. Firms discharging treated industrial wastewaters to the foul sewer pay for the treatment costs imposed on the downstream sewage works.

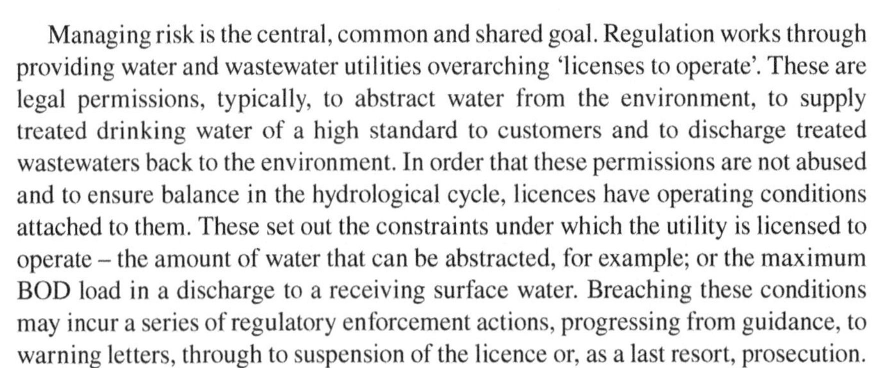

Managing risk is the central, common and shared goal. Regulation works through providing water and wastewater utilities overarching 'licenses to operate'. These are legal permissions, typically, to abstract water from the environment, to supply treated drinking water of a high standard to customers and to discharge treated wastewaters back to the environment. In order that these permissions are not abused and to ensure balance in the hydrological cycle, licences have operating conditions attached to them. These set out the constraints under which the utility is licensed to operate – the amount of water that can be abstracted, for example; or the maximum BOD load in a discharge to a receiving surface water. Breaching these conditions may incur a series of regulatory enforcement actions, progressing from guidance, to warning letters, through to suspension of the licence or, as a last resort, prosecution.

What the regulator seeks above all is confidence, in the public interest, that operations can be managed in a safe and responsible manner and that operators are competent at assessing and managing risk in a preventative way. That is, the operator can preferably anticipate adverse events, act to prevent them and, in the event of an unforeseen incident, act quickly and responsibly to rectify the situation maintaining the trust of its customers. In short, the regulator seeks to work with the utility to develop a demonstrable organisational capability in risk management. With this in place, the regulator, but also investors and the public can have confidence in the operator's delivery of water and wastewater services. Utilities must therefore develop a capability to assess risk, manage it, communicate it and govern it – that is, have an appropriate people and organisational framework in place to ensure that risks deemed unacceptable are acted upon and the residual risk monitored.

Economics plays an important role in regulation. Regulation is usually only appropriate where there is 'market failure', and is achieved through a wide range of instruments compared in terms of their effectiveness and efficiency. In general, 'effectiveness' is the term used to describe how reliably an instrument is expected to bring about the intended environmental objective or outcome. The term 'efficiency' is used broadly to describe how well people use resources to deliver the beneficial results. Economists provide tighter definitions of the related concepts of economic efficiency and cost-effectiveness. Economic efficiency describes the extent to which the distribution of resources in an economy maximises the overall benefits to society. In the context of environmental regulation, policy makers may seek to identify the level of a given pollutant that is economically efficient, reducing pollution to avoid harm and improve the welfare of people adversely affected, while imposing the abatement costs on polluters. Cost-effectiveness describes the extent to which the overall costs to society of achieving a given objective are minimised.

Having identified a target pollution level, say, different regulatory approaches may prove more or less cost-effective in achieving that goal and so there is an increasing interest in the full range of measures and incentives – 'carrots' and 'sticks' – that can be applied to effect the desired outcomes for a range of stakeholders (Figure 5.1). Given this, policy makers and regulators must become adept at decision-making under conditions of considerable uncertainty, and be able to select approaches to policy development and regulatory design (Table 5.1) that take this into account, whether through direct command and control; the use of economic instruments; or the application of modern civic and self-regulation.

EXAMPLE 5.1 WHICH REGULATORY INTERVENTION FOR WHICH TYPE OF HAZARD?

Consider the types of hazard associated with water and wastewater unit operations – both point source and area source releases. Which of the regulatory interventions in Table 5.1 are best suited to point discharges and why?

Next consider the full variety of activities in a catchment that might impact on raw water quality or river/groundwater quality. Design a table of regulatory instruments for a range of activities.

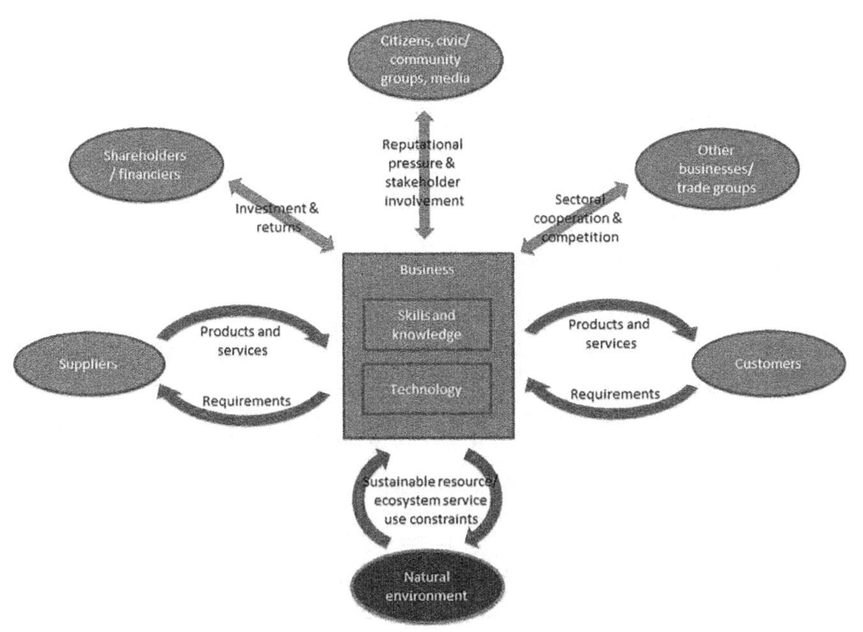

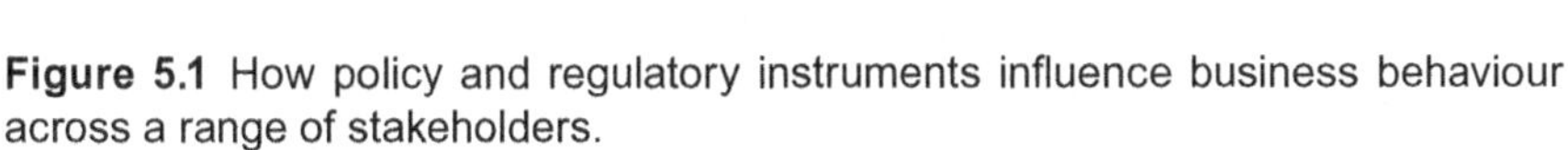
Figure 5.1 How policy and regulatory instruments influence business behaviour across a range of stakeholders.

REGULATORY STYLES OF INTERVENTION

In general, direct regulation is expected to provide the strongest controls on environmental risk, but is also costly to the regulated and government. It can provide a 'level playing field' between competing firms, and tackle illegal behaviours. Direct regulation should be targeted at the highest risk activities, to provide the strongest controls where they are needed most. Other forms of regulation generally provide weaker controls and therefore are suitable for tackling smaller risks or to enhance the effectiveness of direct regulation.

Table 5.1 Three generations of regulatory design.

DIRECT "COMMAND AND CONTROL" REGULATION	
Technology controls	Requirements for businesses to use specific technologies in their operations or products
Zoning/location controls	Performance requirements linked to a specific geography to locate polluters away from sensitive ecosystems
Non-transferable emission licences	Licence to operate according to environmental performance requirements, with compliance monitored and penalties enforced. Inspection requirements can be reduced for good performers through 'earned recognition'
Ambient pollution requirements	Specify required maximum levels of ambient pollution, allowing flexibility to polluters to decide how to achieve
Input restrictions and output quotas	Restrictions are applied in the use or output of products/resources
ECONOMIC INSTRUMENTS	
Payments	Conditional payments made to incentivize a particular activity e.g. provision of ecosystem services
Taxes and subsidies	Change the market price of a good or service, changing the quantity supplied in the market
Tradable rights	Specify a capped quantity of allowances, e.g. to abstract water or to emit carbon, which can then be traded among users
CIVIC AND SELF-REGULATION	
Voluntary regulation	A group of actors agree standards to which individual businesses can sign up
Civic regulation	Community or pressure groups agree performance standards with particular firms
Regulation by professions	A professional body applies standards through conditions of membership
Private corporate regulation	One firm defines standards with which suppliers are required to comply in order to maintain business
Self-regulation	Businesses independently adopt environmental standards, unilaterally or with external verification

Formalised risk assessment and management features widely within regulation because (a) of the need to identify unacceptable risks for onward management; (b) the need for a diagnostic analysis of the practical means by which such risks should be managed (source-pathway-receptor; engineering, organisational and human controls) to acceptable levels; and (c) the need for prioritising risk management efforts by reference to efficiency and effectiveness criteria above.

Preventative risk management underpins the operational delivery of safe drinking water by water utilities, the formalised derivation of water quality guidelines by national and international bodies and the derivation of discharge consent limits by regulators. These are relevant because there is an ever-increasing move towards risk-based regulation, with an expectation that utilities and regulators can both demonstrate their ability to assess and manage priority risks, directing resources to these issues accordingly, and in doing so demonstrate value for money in their risk management activities.

5.2 RISK ASSESSMENT FOR DRINKING WATER GUIDELINES

A long-standing use of risk assessment in water quality management is its application for deriving drinking water quality guidelines. Here we are concerned with managing toxicological (including pathogenic) risks. Raw water sources are exposed to all manner of chemicals and naturally occurring organisms that pose a potential risk to consumers were exposure to occur. Contamination arises from sources including naturally occurring microbial infestation in catchments, the shedding of pathogens by grazing animals, industrial waste disposal (organic and inorganic chemicals), by-products of agricultural practices (e.g. fertilisers, insecticides, pesticides, sediment loss from sloping ground *etc.*) and leisure activities (e.g. petroleum products). Any microorganisms or chemicals present in the raw water supply will be present in the process water and thereby subject to the unit processes of water processing. We use risk assessment to identify the guideline concentrations of these agents that utilities should treat raw water to, in order to protect public health when potable drinking water is supplied.

SOURCES OF SUPPORT ON WATERBORNE PATHOGENS

Pathogens, and the waterborne disease (https://en.wikipedia.org/wiki/Waterborne_diseases) that can result from exposure to them, is a principal health risk that water quality regulations are designed to guard against. Utility risk managers should familiarize themselves with the key pathogens and their relative potencies and fate through water treatment.

(1) The World Health Organization's portal on water sanitation and health (http://www.who.int/water_sanitation_health/diseases/en/) provides an overview.
(2) The US Center for Disease Control and Prevention (http://www.cdc.gov/healthywater/) provides an authoritative summary of the water, sanitation and hygiene agenda (WASH) and key microorganisms of concern.
(3) The Waterborne Pathogens website (http://waterbornepathogens.susana.org/) is a useful source of introductory level information for non-specialists.

The toxicity of the various species in raw water has an important bearing on the water treatment process. How should standards be set for water quality at the tap? Quantitative risk assessment approaches that use a combination of baseline toxicology, exposure assessment and expert consensus have been widely used. They harbour many uncertainties and their application has been controversial. The water quality guidelines produced in this way can only ever be that – guidelines – and professional judgement and pragmatism are required in applying them.

The first step is usually to assess the toxicity of a hazardous agent (chemical or pathogen) and this is often based on animal experiments. Typically, toxicologists quote an LD_{50}, the lethal dose required to kill 50% of the population of animals tested, usually through a defined exposure pathway (e.g. oral, ingestion). This is a single numerical value derived from experimental animal studies in the laboratory that gives no information on the variance of tolerance to different doses, nor a quantitative estimate of the risk of infection from a particular water supply. In general there are then two approaches to setting risk-based standards, namely:

(1) To establish a maximum contaminant level, either as a Guideline (MCLG) or an enforceable Limit (MCL). The risk then becomes the likelihood (probability) of an operator delivering water of a quality that exceeds the technical limit (consequence). Because there is usually a high degree of conservatism, or in-built safety in deriving these values, exceeding this limit on occasion may not necessarily lead to a health detriment or adverse impact. Exceedance should, however, be a trigger for action.

(2) The second approach is to consider the probability of exceeding the level of contaminant that produces a health risk. This is a measure of the true risk experienced by the company and the consumer.

Assessments can be made of risks associated with all species that have been identified to exhibit potent toxic behaviour. The task during risk assessment is to estimate the frequency (or probability) of events with adverse consequences to the consumer. The analysis requires several pieces of information for each and every possible toxic species and especially:

- a dose-response assessment (a characterisation of the relationship between an absorbed dose and an adverse response); and
- an exposure assessment (an estimate of the dose at the point of exposure).

5.2.1 The dose-response assessment

The first step of the risk assessment is to determine the relationship between the amount or duration of exposure to the toxic agent under consideration and the extent of injury, or adverse response that may occur. Even where good epidemiological studies exist, reliable quantitative data in humans is rarely available. In most cases, dose-response relationships must be predicted using tests undertaken on experimental animals. This raises three important limitations that introduce substantial uncertainties in the assessment of risk:

- Animals are exposed at very high dose levels and so effects at low doses must be determined by extrapolation.
- Animals and humans differ in their susceptibilities and hence their response to contaminants.
- The human population is heterogeneous. Therefore, they have varying 'within-population' degrees of susceptibility. Infants, older people or populations with poor nutrition may be more vulnerable than fit, healthy and well fed populations.

In characterising dose-response relationships, toxicologists usually distinguish between two types of substance behaviour:

- substances that exhibit threshold effects; and
- substances that exhibit no-threshold effects.

It is widely assumed that, for non-carcinogenic substances, there is a limiting dose, below which there are no observed adverse effects (a no observed effects level or NOAEL). This has been interpreted to mean that there exists a dose of a toxic substance below which the risk to health is zero. However, assuming the threshold exists, strictly speaking zero risk is only possible if the level of exposure can be guaranteed to be below the threshold. There is usually some finite probability that the dose received by the population will exceed the threshold. Some studies are unable to define a level at which there is no observed adverse effects. In these cases, it may be possible to determine a level at which there is a lowest observed effect (LOAEL).

The NOAEL values, after scaling for body weight, have been used to set standard reference doses (R_fD) for drinking water by the World Health Organisation (WHO). It is accepted that these threshold limits are highly uncertain and it is usual to apply uncertainty (or safety) factors of varying orders of magnitude to the experimental data, (e.g. NOAEL/100). The greater the uncertainty, the greater this factor. Factors of 10, 100, 1000 or 10,000 have been applied in practice for different substances. The scientific justification for the selection of these apparently arbitrary factors is somewhat limited.

At present, carcinogenic substances are treated as having no threshold for a toxic effect. This implies that zero risk of cancers exist only at zero dose. As the dose increases above zero, the risk becomes finite and increases as the dose increases. As our understanding of carcinogenesis as a multi-stage process increases, however, the premise of a non-threshold approach for these agents is being challenged.

Regulators often use the concept of a maximum allowable contaminant levels (MACLs) to ensure that health risks are minimised. This is sometimes referred to as the drinking water equivalent limit, or DWEL. The main sources of information on which maximum contaminant levels are set are epidemiological studies and dose response studies on animals. Dose response studies are the greatest source of information. The first step in deriving these criteria is to determine the NOAEL or LOAEL, usually from dose response studies on animals. The NOAEL or LOAEL is then divided by an uncertainty factor (UF) of 10, 100, 1000 or 10,000 depending on the quality of the data on which the assessment is based, to generate the reference dose R_fD.

$$R_fD = \frac{\text{NOAEL}}{\text{UF}} \quad \text{or} \quad R_fD = \frac{\text{LOAEL}}{\text{UF}} \tag{5.1}$$

For determining lifetime health advisories (guidelines), a drinking water equivalent limit is determined by multiplying the R_fD by the average weight of a human (e.g. 70 kg) and dividing by a human water consumption rate, typically 2 l/day.

$$\text{MCLG} = \text{DWEL}\,(\text{mg/l}) = \frac{\text{R}_\text{f}\text{D}\,(\text{mg/kg/d}) \times \text{W}\,(\text{kg})}{V\,(\text{l/d})} \tag{5.2}$$

This may then be adjusted for lifetime use. For protected individuals such as children, health advisories assume more stringent criteria e.g. assume 10 kg body weight and a water consumption rate of 1l for children. As an illustrative example, consider the herbicide atrazine. A number of toxicity studies have been carried out. The most appropriate data related to a study using dogs.

$$R_fD = \frac{0.48\,(\text{mg/kg/d})}{100} = 0.005\,\text{mg/kg/day (after rounding up)}$$

A NOAEL of 0.48 mg/kg/d is based on an absence of cardiac pathology or any other adverse effects in dogs; 100 is the uncertainty factor chosen in accordance with EPA guidelines.

$$\text{DWEL} = \frac{0.0048\,(\text{mg/kg/day}) \times 70\,(\text{kg})}{2\,(\text{l/day})} = 0.168\,\text{mg/l}$$

$$= 200\,\mu\text{g/l (after rounding up)}$$

where 0.0048 is the R_fD before rounding up; 70 kg the mean body weight of an adult human; and 2 l/d, the assumed daily water consumption of an adult.

$$\text{Lifetime health advisory} = \frac{0.168\,(\text{mg/l}) \times 0.2}{10} = 0.00336\,\text{mg/l} = 3\,\mu\text{g/l}$$

0.168 is the DWEL before rounding up; 0.2 the assumed relative contribution of atrazine from drinking water exposure (20%); and 10 is an additional uncertainty factor to account for possible carcinogenicity.

In USEPA advisory documents, the term margin of exposure is used for non-carcinogens and for the non-carcinogenic effects of carcinogens. MOE is defined as the ratio:

$$\text{MOE} = \frac{\text{NOAEL}}{D} \tag{5.3}$$

where NOAEL is the level at which there is no measurable adverse effect and D is the exposure dose.

Credit: istockphoto © vitranc

The health advisory programme sponsored by the USEPA's Office of Drinking Water provides information on health effects, analytical methods and treatment technologies that may be useful in dealing with contaminated drinking water. These refer to non regulatory contaminants and define safe drinking water contaminant limits with a margin of safety to protect sensitive members of the community. Health advisories contain a range of useful information to help with risk analysis.

The level of human tolerance to chemicals or microorganisms is usually estimated using animals to model the human response. For carcinogenic substances, the approach is to administer the substances under investigation to a population of animals at different doses. In this case, it is usually assumed that there is no threshold for infection. The proportion of animals that die from cancers at a particular dose is plotted as a dose-response function. Historically, in order to produce meaningful results in experimental animals, high dose rates were administered to small animals (e.g. rats or rabbits) with relatively short natural lives. The analyst fitted the dose-response data to a mathematical function and extrapolated the results to some target level of probability of fatality from cancer. The extrapolation functions can give highly variable results. The functions employed include the:

- one hit;
- multi-stage;
- multi-hit;
- Weibull; and
- probit models.

The fitted multi-stage method gives conservative results (the USEPA uses a linear multi-stage model). In order to estimate a dose it is necessary to define a level of acceptable probability, or an 'excess lifetime risk' over and above the background risk. If the tolerable probability of death is accepted as 10^{-5}, then the corresponding target dose for the animal is 0.1 mg/kg · d. This is the dose that brings about an increased excess, (above background), likelihood of 10^{-5} of a fatal cancer in animals. This has then to be scaled up to accommodate the difference between the animal model and humans. This has been carried out using a weight ratio method, using the tolerable human toxic dose (D^*) as:

$$D^* = \text{tolerable animal toxic dose} \times \{W_h / W_a\}^{3/4} \qquad (5.4)$$

Where W_h is the weight of humans and W_a is the weight of the test animal, (usually these calculations are based on averages).

The standard approach assumes an animal is a suitable model for a human and that size, (weight ratio to a power), is the only factor of importance in scaling the dose limits from animals to humans. Clearly, uncertainties on dose tolerance are large in the absence of other evidence on human behaviour. Similar animal studies are carried out to assess non carcinogenic chemical and microbial species. All such tests are capable of generating dose-response curves that can form the basis of a quantified risk assessment, though given the substantive uncertainties, caution and intelligent interpretation are both required in managing these data.

5.2.2 The exposure assessment

The actual dose, D, of a particular toxic substance that a human will consume depends on the ingress route into the body and the following factors:

- the concentration of the substance in water, × (mg/l);
- the volume consumption rate of water, V (l/d) (combining all routes);
- the fraction, f, of the substance absorbed by the body; and
- the body weight, W (kg) (average weights are normally assumed).

$$D = \frac{f \times V}{W} \text{(mg/kg} \cdot \text{d)} \qquad (5.5)$$

Where f is the fraction of material absorbed and D is the absorbed dose.

In practice, there will be large uncertainties in the actual dose and it could also be highly variable because of changes in drinking patterns over a long period of time.

5.2.3 Preventative water quality risk management

Above we have summarised the tenets of regulatory risk assessment as applied to the derivation of water quality standards or guidelines – those limits that drinking water utilities are expected to operate to in order to supply safe drinking water. However, we recognise the limitations of compliance alone as a basis for managing risk. Usually by the time the supply is identified as being out of compliance, exposure will have already occurred; hence the need for a different philosophy or approach for drinking water supply – a preventative approach that focusses on total system from catchment to tap, deployed with an organisational culture of vigilance, or 'mindfulness'.

The control of drinking water risks is largely, though not exclusively, in the hands of the water companies. The risk of disease, infection or fatality, arising from the presence of microbial organisms is routinely reduced by the use of filtration and water disinfection processes, such as chlorination. It is widely accepted that water quality monitoring by reference to risk-based standards, (contaminant concentrations,) and operational controls, (e.g. maintenance of a chlorine residual), is an essential component of the risk management strategy of a water utility.

However, it is not usually possible to quantify the benefits of monitoring on risk management. The efficiency and effectiveness of chlorination will not be the same for all microbial species and all water qualities (e.g. microorganisms may be able to use solid surfaces to protect themselves from the hypochlorite). Even if a complete detection method was developed, there remains the significant problem of response time. Unless the toxicity of the species had been studied prior to identification in the water supply, there would be no data available on which to assess the possible consequences in time to influence management decisions.

It should be recognised that while it is possible, in principle at least, to assess and manage the risks associated with known chemicals and microbial species, there are some fundamental limitations to risk analysis at the hazard identification stage. Firstly, there is no way of knowing, in advance, what new species are likely to be present in water sources and since a general microbial or chemical detection system capable of identifying any new species does not currently exist, there will always be some degree of unquantifiable risk. Some authorities suggest the total number of potentially pathogenic microorganisms is unknown and increasing e.g. *Crytosporidium parvum* was not recognised as a human pathogen until the mid-1970s.

In recognition of the need to adopt a preventative risk management approach to drinking water quality management, the Wold Health Organisation introduced the water safety plan (WSP) approach in the third revision of its drinking water guidelines in 2004 and strengthened these in 2011. Building on the Australian guidelines for the management of drinking water quality (NHMRC, 2004), the objectives of a WSP are to prevent contamination of raw water sources, treat water to remove contamination and prevent re-contamination during storage distribution and handling.

The primary aim is to protect public health through system assessment, operational monitoring and management plans (Figure 5.2); guided by health-based targets and overseen by surveillance (Davison *et al.* 2005). One aspect that most WSP guidance and case studies agree on is that 'buy-in' from across the organisation, and particularly from senior management, is imperative to successful implementation (WHO, 2004; IWA, 2004; NHMRC, 2004; Godfrey & Howard, 2005).

RISK MANAGEMENT FOR ASSURING SAFE DRINKING WATER

Hrudey *et al.* (2006) suggest elements that water utilities may wish to consider when trying to develop mindfulness:

- Informed vigilance actively promoted and rewarded.
- An understanding of the entire system, its challenges and limitations are promoted and actively maintained.
- Effective, real-time treatment process control, based on understanding critical capabilities and limitations of the technology, is the basic operating approach.
- Fail-safe multi-barriers are actively identified and maintained at a level appropriate to the challenges facing the system.
- Close calls (near misses) are documented and used to train staff about how the system responded under stress and to identify what measures are needed to make such events less likely in the future.
- Operators, supervisors, lab personnel and management all understand that they are entrusted with protecting the public's health and are committed to honouring that responsibility above all else.
- Operational personnel are afforded the status, training and remuneration commensurate with their responsibilities as guardians of the public's health.
- Response capability and communication are improved.
- An overall continuous improvement, total quality management mentality pervades the organisation.

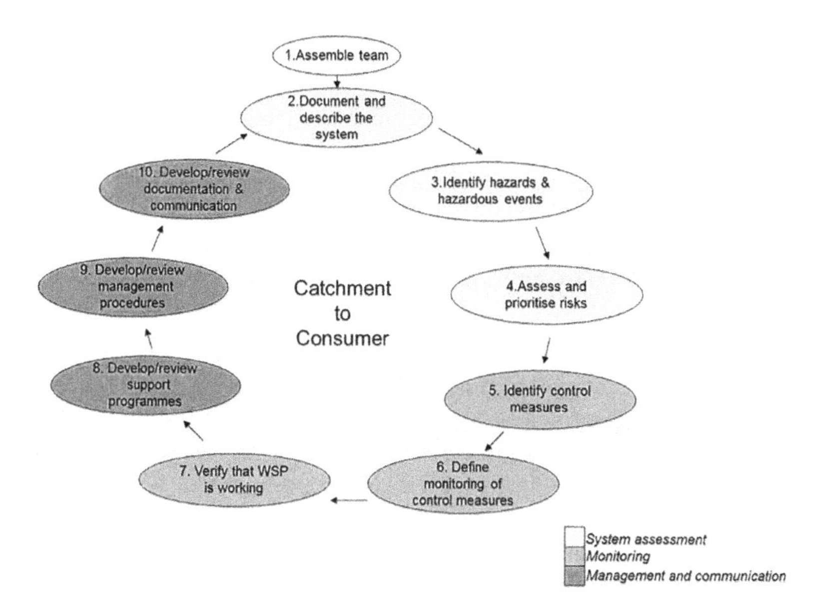

Figure 5.2 Example 10 step process for the development of a water safety plan.

WATER SAFETY PLAN CONTROL (RISK MANAGEMENT) MEASURES

Control measures (step 5 in Figure 5.2) are those steps in supply that directly affect water quality and which, collectively, ensure that water consistently meets health based targets. They are actions, activities and processes applied to prevent or minimise hazards occurring. Control measures work by:

- reducing the entry of contamination into the water supply (e.g. run-off interception); and
- reducing the concentration e.g. coagulation/flocculation.

Applying the principles of preventative risk management to the well-established multiple-barrier approach, successful implementation of a WSP requires utilities to understand their drinking water supply system; prioritise the risks; and design appropriate control measures.

The regulation of water quality risk by virtue of the WSP approach is an example of civic and self-regulation. By and large, adoption of WSPs has been through light touch regulation and through the momentum generated by an industry 'code of practice', facilitated through the World Health Organisation and International Water Association. The development of supporting guidance, of case study WSPs, of training material and of a web-based information portal have all contributed to their successful 'roll-out' in the sector.

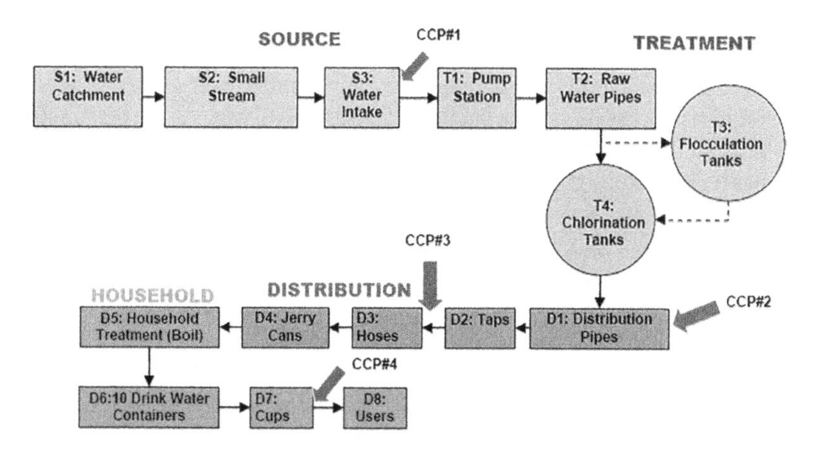

Figure 5.3 Example critical control points (CCP#1, #2 and #3) identified within a system from catchment to tap.

5.3 MANAGING RISKS OF ABSTRACTION AND DROUGHT

5.3.1 Licensing abstractions

Having considered the underlying basis for water quality standards and their management through the adoption of water safety plans, we now consider the management of risk within the broader water cycle. Here we are concerned with the amount of water abstracted and the likelihood and consequences of over abstraction on other uses, including ecosystems. A regulator's responsibility for managing water resources typically encompasses:

- the conservation, redistribution, or augmentation of water resources;
- securing of their proper use;
- powers to grant, vary or refuse water abstraction and impoundment licences (or permits) on application;
- powers to revoke or vary existing licences; and
- powers to monitor and enforce abstraction and impoundment licence conditions.

The successful management of water resources involves permitting abstraction to the extent necessary to meet an abstractor's foreseeable and reasonable needs, whilst ensuring the resource and the water-dependent ecosystem, together with existing protected rights and lawful uses, are safeguarded. Water is abstracted from rivers, canals and aquifers for a range of uses including the public water supply, irrigation for agriculture and use as process water within industry. The regulator's principal role within respect to licensing these abstractions is to balance the competing demands for water from these users with the needs of the environment through an abstraction licensing system.

An abstraction licence gives the holder of the licence a right to take water from a stated source for the stated purpose at stated rates, subject to conditions specified, until it expires or the holder or the regulator revokes it. The licence does not guarantee the quality of the water is suitable for its intended purpose, nor that the amount licensed will always be available. In submitting an application for a licence, the potential abstractor usually undertakes a resource assessment. The potential risks identified in this assessment, and therefore posed by the new abstraction, are managed by the regulator using licence conditions. In England and Wales these may include conditions for protecting the aquatic environment, other users and to ensure the proper management of water resources:

Credit: istockphoto © DutchlightNetherlands

- **Time limit**. All new licences and variations, (other than downward variations or other similar minor variations having no environmental impact), have a time limit imposed.
- **'Hands-off' flow or level conditions**. In order to protect low flows and flow sensitive ecosystems, flow or level constraints can be applied to abstraction licences, principally those authorising abstraction from surface waters. This requires a cessation or reduction of abstraction when a particular flow or level in the source or nearby monitoring point is reached.
- **Metering**. All significant abstractions are required to be measured by an appropriate meter or other suitable device. The details of the meter readings (monthly, weekly or daily – as specified in the licence) are provided to the regulator. The data is used to assess the demand, actual abstraction and resource assessments – either as part of the catchment management process or with regards to specific areas or reaches.
- **Return of water after use**. In some instances, the return requirements after use will be specified to limit the net loss of water from the source.
- **Water quality**. Where the abstraction may have an impact on water quality (e.g. groundwater abstraction may cause movement of pollutants or allow saline intrusion in coastal areas), quality parameters may be defined leading to a reduction or cessation in abstraction.
- **Monitoring requirements**. Special monitoring provisions may be included to protect a nearby water-dependent environment (e.g. a sensitive wetland; salmon spawning grounds), from the potential effects of the abstraction or to provide further information on the environmental impacts of the abstraction.

5.3.2 Droughts

Droughts are natural events. In the absence of significant rainfall, groundwater levels, spring discharges and river flows fall. We cannot influence the likelihood of droughts occurring but we can manage their consequences. Competition for water resources, especially during times of exceptional drought, requires a balanced approach towards water supply, environmental and other interests.

Drought management is concerned with managing needs and demands for water during droughts. Water utilities and other abstractors are usually expected, by the regulator, to plan and manage their need for water within the constraints of their abstraction licences. In severe droughts, however, utilities may temporarily need to abstract outside the terms of their licence. For exceptional circumstances, 'drought orders' and permits, forms of temporary authorisation to alter the existing abstraction regime, allow these exemplary needs for water to be met. Contingency plans for managing the consequences of severe drought might reasonably be expected to:

- demonstrate that reasonable measures have been taken to reduce demand;
- take action when or where least environmental damage will occur; for example, in the winter;
- locate sites for additional water where as little as possible environmental damage will occur; and
- include measures to mitigate the most serious impacts.

5.4 RISKS AND ASSET MANAGEMENT

We introduced assets early on in this book. They are conventionally regarded as the hardware of water and wastewater treatment and are true 'assets', in a balance sheet sense, because much of the financial value of utilities is tied up within the unit processes and infrastructures used to deliver treated water and wastewater.

When we operate water and wastewater plant, we put these assets at risk (of failure) for the purposes of protecting public and environmental health. Many assets are run continually. Increasingly, utilities 'sweat' (push to the limit) their assets; that is, they run them fully optimised over the full length their design lifetime, and sometimes beyond. This puts the assets under greater strain and inevitably increases the importance of well-designed maintenance, refurbishment and asset renewal programmes which, if to be effective, are best targeted on those assets at greatest risk.

EXAMPLE 5.2 MANAGING INVESTMENT

The water industry in the England and Wales is regulated by the government Office of Water Services (OFWAT). OFWAT has introduced a 5 year Asset Management Period (AMP) and determines the funding that utilities may spend on capital maintenance and investment. A 5-year cycle drives the planning, design and construction programme for all water companies in England and Wales. The cycle has the consequences of:

- synchronized demand and supply patterns for resources, e.g. capital, labour;
- temporary inflation of resources prices;
- temporary underutilization of resources, in particular labour; and
- low profitability with water price capping
- inflated water prices.

The resources required to deliver projects within this programme cycle undergo substantial market forces. Years 1 and 2 are characteristic by a labour demand in design. Years 3 and 4 are characterised by increasing activities in asset creation and project implementation. Avoiding surges in demand on the resource markets reduces the risk of excessive costs to project/AMP programme delivery. However, within this regulatory framework and strategy, the water companies have limited influence on managing interaction with resource markets.

Asset management is the process of designing, commissioning and maintaining the asset base. Risk assessment plays an important role in that, in order to manage the assets effectively, the utility must develop a dynamic view of the likelihood and consequences of asset failure across the portfolio of assets so that both operational expenditure (OPEX) and capital investment (CAPEX) can be prioritised and targeted to where it is most needed. Inherent to this view of asset risk is an understanding of the **condition** of an asset, its **criticality** to the overall function that it undertakes and the **consequences** of a failure in asset performance. Markov models, similar to those introduced in unit 4 have been widely used to estimate the risks of failure and develop a whole-life costing approach to asset management that balance OPEX and CAPEX expenditure on a risk-informed basis.

The new international standard ISO 55000 defines asset management as a coordinated activity of an organisation to realise value from assets, these being any item, thing or entity of value to an organisation. Asset management, to optimise value, requires the balancing of costs, opportunities and risks against the desired performance of an organisation's assets, to achieve strategic objectives objectives. Importantly, systematic asset management allows the application of a series of analytical approaches to securing value over the whole life cycle of an asset from design to final decommissioning. It has been widely adopted in the water utility sector to minimise the whole life cost of assets but also inform risk management and business continuity planning (Table 5.2).

Table 5.2 Perspectives on water utility assets.

Driver	Expectation from Investment	Definition/comment
Regulators	Performance	As monitored under a limited number of performance measures defined by regulators. Measure is backward looking.
Customers	Serviceability	Ability of the assets to satisfy customer expectations – a wider, forward looking definition of serviceability than 'performance'.
Investors and shareholders	Asset value	The focus on serviceability must not be at the expense of asset value, which must be maintained to satisfy shareholder interests.
Competition	Value for money	Achieving the correct balance between investment and risk will achieve stakeholder satisfaction and deter the competition.

Utilities usually have a statutory duty to maintain their assets so that they can provide water and/or wastewater services to current and future customers, and to protect the environment. This 'service' includes providing clean water to customers, pollution control to the environment, and the avoidance of daily disruptions to third parties in the community (e.g. traffic congestion). These services are provided via a variety of physical assets that deteriorate with age and with use and managing the risks of failure is a delicate balance.

If a high risk of failure from deteriorating assets is tolerated and investment deemed to be too low, serviceability and asset value will decline, resulting in the loss of customers and shareholders to competitors. When the risk of asset failure is maintained at too low a level, over-investment leads to a higher level of service, but also higher customer bills that tend to result in customer dissatisfaction and the utility becomes vulnerable to competitive pressures.

The key to sound asset management is securing the right balance of cost, risk and performance across the asset base in a visible and auditable manner, and within statutory, environmental, social and financial constraints. Clearly, if a utility left an asset too long before replacing it, it might fail and the service (supply, business, confidence) be interrupted. Conversely, if the asset were replaced too early, some of its capability would be wasted as would a portion of the finances invested in the asset. One might argue that the ideal time (financially) to replace an asset would be at the instant before its designed end-of-life. Adopting this approach, however, would also mean that the portfolio of assets were being 'sweat' to a point of near failure and the system as a whole might then become more precarious in risk terms – 'teetering on the edge' of failure, as it were. This would be unacceptable as there has to be some built-in 'redundancy' as a margin of safety for systems that deliver public health protection.

Most utilities have thousands of assets the performance of which they are continually seeking to optimise. The key dependencies are a good asset register, standardised operating practices for design, build and operate phases and reliable costs information so that defensible financial investments can be argued for and secured. Risk analysis tool that have been used to assist in utility asset management include:

- risk ranking systems adopted to ranks the condition (likelihood of failure), criticality (importance of the asset to service delivery) and consequences of failure;
- asset surveys, to provide detailed information of the relative performance and condition grade of assets;
- asset renewal modelling of the remaining design life and expected deterioration curve of assets and combinations of assets in series or parallel; and
- reliability-centred maintenance for targeted maintenance scheduling according to risk.

5.5 REGULATING WASTEWATER DISCHARGES AND VOLUNTARY INITIATIVES

To protect the natural environment, the effluent quality of treated wastewater is regulated. Permission has to be secured from the relevant authority to discharge treated wastewaters back to the environment. We are usually concerned with either sewage treatment works owned by water and/or wastewater utilities, or with in-plant wastewater treatment plant (WWTP) discharging to surface water or the foul sewer infrastructure. For example, integrated refineries, pharmaceutical works or large steelworks often have their own industrial wastewater treatment plants that discharge treated wastewaters to sewers or surface waters and therefore require regulation.

In the UK, if discharging to surface waters, groundwaters, estuaries and coastal areas, permission has to be sought from the relevant environment agency. If discharging to a public sewer (trade effluent), permission is required from the receiving water utility whose own assets may be put at risk from the hazardous constituents in a trade effluent discharge. The contribution that sewage works and industrial discharges make to the flow and load of rivers (Table 5.3) means that regulatory control on the quantity and quality of the discharges is essential to control and improve river water quality.

Regulators issuing consents for discharge typically require information about (i) the type of effluent to be discharged; (ii) the volume to be discharged; (iii) the constituents of the effluent; (iv) where it will be discharged and where the discharge points lead to; (v) whether the effluent will be treated and how; and (vi) where the effluent can be sampled.

Hazardous constituents typically include the following when they are toxic, persistent and liable to bio-accumulate:

(a) organohalogen compounds and substances which may form such compounds in the aquatic environment;
(b) organophosphorus compounds;
(c) organotin compounds;

THE EC WATER FRAMEWORK DIRECTIVE

The Water Framework Directive (WFD) is a significant piece of water resource legislation. The Directive offers a holistic approach to water management and aligns EC water legislation through introduction of a statutory system of analysis and planning based upon the unit of the 'river basin'. The objective of this legislation is to achieve 'good status' for all European water bodies by 2015. Specifically, the Directive aims to:

- prevent deterioration and enhance the status of aquatic ecosystems and associated wetlands;
- promote the sustainable consumption of water;
- reduce pollution of waters from priority substances;
- prevent the deterioration in the status;
- progressively reduce pollution of groundwaters; and
- contribute to mitigating the effects of floods and droughts.

The water industry has a variety of interfaces with the Water Framework Directive. Water quality parameters, in particular, wastewater discharge consents but also issues around water abstraction and consumption are going to affect the water companies in Europe.

(d) substances and preparations, or the breakdown products of such, which have been proved to possess carcinogenic or mutagenic properties or properties which may affect steroidogenic, thyroid, reproduction or other endocrinerelated functions in or via the aquatic environment;
(e) persistent hydrocarbons and persistent and bioaccumulable organic toxic substances;
(f) cyanides;
(g) metals (in particular cadmium and mercury) and their compounds;
(h) arsenic and its compounds; and
(i) biocides and plant protection products.

Table 5.3 Illustrative historic contributions to point and diffuse sources of pollution.

Location	Determinant	Contribution
Wharfe to Boston Spa	Nitrate	Point sources 152×10^3 kg/yr (17%); diffuse sources 764×10^3 kg/yr (83%); 655×10^3 kg/yr exits catchment so losses through denitrification are 261×10^3 kg/yr.
Thames to Teddington	Orthophosphate	7683 kg/d from WWTPs (90%), 872 kg/d from agriculture (10%)
Thames to Teddington	Flow	40% of total river flow from WWTPs in most summers but 73% during droughts
Severn-Trent	Flow	Mean percentage of total river flow that is sewage effluent varies from close to zero to 34%.
Entire aquatic environment of EC	Phosphate	Sewage and industry contribute 41% of total phosphate

Rivers can differ widely in their contribution of treated wastewaters, from those with approximately 100% of emissions from diffuse sources in remote agricultural areas, to those with close to 100% from point-sources in urban, highly industrialised areas.

The principal means of control over water quality is the discharge consent. Discharges are 'allowed' by the regulator; that is, they are consented to, with conditions attached. Effluent consents are informed by river quality modelling that estimates the loads entering the river and assesses the potential impacts of these loads on the quality of the receiving water body. Many small discharges to the aquatic environment have a low potential to adversely affect the receiving water because of the nature and low volume of the discharge and would be difficult to control by means of a numerical consent. In these cases, descriptive consents are applied which can typically define the nature of the effluent treatment plant to be used, plus a requirement that the plant be correctly operated and adequately maintained. Many non-numeric consented discharges are only authorised to operate under specific weather conditions – for example, under storm conditions.

In contrast, a sewage treatment works' numeric consent can include limits for a range of substances, but the three principal consented limits are for suspended solids (SS), biochemical oxygen demand (BOD) and ammonia. When a discharge fails to comply with its consent limits and conditions it has 'breached' the consent. Breaches of consent conditions in all forms of discharges are often relatively minor, although all such breaches need to be addressed. Very often, the "20/30 standard" is quoted for the discharge of sewage treatment works effluents to the environment. This refers to 20 mg/l BOD and 30 mg/l suspended solids. The standard was adopted following the 1912 Royal Commission on Sewage Disposal when the 5-day BOD test was made standard. It is based on an 8:1 dilution by the receiving water, assuming that receiving water's BOD is <4 mg/l.

Industrial discharges are normally controlled by absolute limits on concentration and flow. In some cases, an absolute limit on load discharged may also be required. In this situation, the basis for the load calculation will be specified, as different estimates will be obtained from the multiplication of an instantaneous flow and concentration or a 24 hour composite sample concentration multiplied by the daily mean flow. The mass balance equation is the simplest way of expressing the instantaneous complete mixing of an effluent stream with receiving water. This takes the form:

$$C_1Q_1 + C_EQ_E = C_2Q_2 \tag{5.6}$$

Where

C_1, Q_1 = concentration and flow above the point of discharge
C_E, Q_E = concentration and flow of effluent
C_2, Q_2 = concentration and flow after the point of mixing of the discharge and the receiving water

Although not all consents are to the '20/30' standard, many sewage treatment works' discharges have conditions for BOD and suspended solids. With new consents for nitrate and phosphorus, particularly for large sites discharging directly into major rivers, ever tighter limits are in place for ammonia, orthophosphate and BOD. In addition, consents may be required for groups of compounds, such as endocrine disrupting compounds (EDCs). Meeting consents set for ever increasing water quality has obvious cost implications.

The discharge of effluent to sewer by industry is known as 'trade effluent' and requires permission before discharge. The advantages of a sewer discharge are that it provides flexibility, a reliable means of wastewater disposal for small traders, opportunities for dilution, and that they allow treatment at a utility's wastewater treatment plant (WWTP). The limitations relate to the need to ensure safe conveyance to the sewer, the invisibility and thus limited possibility for spill detection and, importantly, the potential for toxic shock loads at the WWTP with the potential for catastrophic loss of treatment, usually of secondary biological treatment (e.g. cyanides, phenols). Trade effluent consents are issued for the purpose of managing a series of risks (Table 5.4).

Table 5.4 Risks managed through trade effluent control.

Harm avoided	Provide data for
– Harm to process	– Process management
– Unacceptable storm discharges	– Process planning
– Breach of legislation	– Treatment charges
– Harm to staff	
– Corrosion of the sewer fabric	
– Overloading and flooding of properties	
– Blockage	
– Formation of explosive, inflammable and toxic gases	

The behaviour of organisations and individuals has a marked influence on the likelihood of accidents or unwanted consequences occurring. In a regulatory climate where many of the point sources of pollution have been brought under control, regulatory attention has begun to focus on more diffuse sources of pollution – the over-application of nitrogen fertilizers to fields and pesticide use, for example. Risk management in these settings is more challenging. Rather than dealing with closed engineered systems (wastewater treatment plants, oil-water interceptors, trade effluent discharge points; unit processes with standard operating procedures) it is more obviously the behaviour (motives, level of environmental awareness, duty of care) of the discharger that determines how these risks to the environment are managed. Here, education and the dissemination and uptake of codes of good practice become more powerful vehicles for managing risk. Much of the regulatory effort here is in drafting and promoting best practice with specific sector groups for mutual advantage. Several good examples of these partnerships exist:

- the 'safe sludge matrix' was developed by the UK water industry in concert with the regulatory authorities and agriculture sector guide farmers on the application of treated sewage sludge (biosolids) to land (the risks of sewage sludge derived pathogens entering the food chain);
- the PEPFAA code (Code of Prevention of Environmental Pollution from Agricultural Activity) was developed in Scotland in consultation with water authorities, land users and managers, the Scottish Executive and the Scottish Environment Protection Agency (SEPA) (the risks to water quality of diffuse agricultural pollution);
- the Oil Care Code and Campaign is a regulatory initiative of the Environment Agency of England and Wales to reduce oil pollution by providing guidance on and facilities for the safe disposal and management of oil (the risks to water courses of shock loads from oil pollution events, especially those resulting from losses from unbunded oil tanks); and
- the Sustainable Urban Drainage Scottish Working Party (SUDSWP) was established by local authorities, the water utility sector, Scottish Enterprise and developers, and paved the way for more sustainable drainage in Scotland (the risks to surface and ground waters from polluting urban drainage).

Credit: istockphoto © RuudMorijn

5.6 THE REGULATION OF RISK MANAGEMENT

The increasing adoption of risk management at the organisational, or enterprise level, also raises issues of regulatory compliance for the 'internal audit' function of water utilities. All organizations face risks to their strategic, tactical and operational objectives; water utilities are no exception. Progressive utilities recognize their business environment is dynamic and uncertain, requiring forward looking, proactive risk management. They recognize that seizing business opportunities necessitates accepting some amount of risk and that core to this is effective and well-informed risk governance. Having this in place is the foundation of an effective relationship with the regulator.

REGULATING RISK GOVERNANCE

In the UK, Board level risk governance is specified by the 2014 Financial Reporting Council guide (FRC, 2014). The Board is responsible for determining the nature and extent of the principal risks it is willing to take in achieving its strategic objectives. The Board should maintain sound risk management and internal control systems.

Directors should confirm in the annual report they have carried out a robust assessment of the principal risks facing the company, including those that threaten its business model, future performance, solvency or liquidity. Directors should describe those risks and explain how they are being managed or mitigated.

There are a number of established risk governance frameworks. The most widely used are the International Standards Organisation ISO 31000 and Committee of Sponsoring Organizations of the Treadway Commission (COSO) frameworks. A key driver for implementing these has been the apparent failure of the Boards of financial institutions to have adequate understanding, oversight and control of their institutional risk taking. The core business risk assessment and management processes for organisations are set out in the international risk management standard ISO 31000:2009. Identifying, analysing and evaluating risk forms the basis of good risk governance. These processes are increasingly embedded in the business of water utilities, from drinking water quality management to asset management, and therefore become part of 'business as usual'. As risk management processes mature in the water sector, they will become reward driven and aligned with ambitions for reputational enhancement, competitive advantage and continuous improvement. Leadership support remains key to gaining the buy-in required to embed risk management processes within organisations.

Many water utilities now possess a formalised risk governance function with in-house expertise, whether in 'City Hall', or as part of the head office strategic function, or else distributed within their businesses, for example in asset management. The true value of risk management comes when it is integrated with other business functions to support business value creation through resilient business planning; efficient asset management; heightened emergency preparedness; better access to finance; avoiding breaches of compliance; reduced business losses, and reduced insurance premiums and litigation penalties.

Forward-looking utilities are now looking to how their risk management capabilities can better support other strategic functions. Integrating risk into the core business functions of water utilities is necessary to generate the authentic business value and strategic opportunities of risk management. This is the focus of our next chapter.

5.7 SUMMARY AND SELF-ASSESSMENT QUESTIONS

We have considered the regulatory perspective on water and wastewater treatment and the role of risk analysis and management in controlling releases and securing safe drinking water. Modern regulation is increasingly concerned with targeted regulatory responses to managing key hazards at source and in securing the safe and responsible management of water utility assets (including their replacement) at a reasonable price for the customer. Modern legislation is equally concerned with activities in catchments and with the shared responsibilities of the many actors whose activities have an impact on water quality. Since the 1970s, regulation has progressed beyond what could historically be viewed as an adversarial approach to one of partnership.

Next, we consider the business risk management activities of utilities. This extends our discussion to a consideration of how utilities manage the multitude of corporate threats and opportunities in their day to day operation.

SAQ 5.1 Consider yourself to be a regulator of a large sewage treatment works. What specific evidence would you seek as assurance of good risk management? Other than consent records, what would you look for during an inspection of the works?

SAQ 5.2 What risk management advice would you offer farmers to prevent spray drift to surface water bodies. Include advice that addresses both the probability and consequences of undesirable ecological impacts.

5.8 FURTHER READING

Abbott M. and Cohen B. (2009). Productivity and efficiency in the water industry. *Utilities Policy*, **17**, 233–244.

Allan R., Jeffrey P., Clarke M. and Pollard S. (2013). The impact of regulation, ownership and business culture on managing corporate risk within the water industry. *Water Policy*, **15**, 458–478.

Bel G. and Warner M. (2008). Does privatization of solid waste and water services reduce costs? A review of empirical studies. *Resources, Conservation and Recycling*, **52**, 1337–1348.

Byleveld P. M., Deere D. and Davison A. (2008). Water safety plans: Planning for adverse events and communicating with consumers. *Journal of Water and Health*, **6** (Suppl.1) 1–9.

Committee of Sponsoring Organizations of the Treadway Commission (2004). Enterprise Risk Management Integrated Framework. Executive Available at: http://www.coso.org/documents/coso_erm_executivesummary.pdf.

Dalgleish F. and Cooper B. J. (2005). Risk management: Developing a framework for a water authority. *Management of Environmental Quality: An International J.*, **16**, 235–249.

Davison A., Howard G., Stevens M., Callan P., Fewtrell L., Deere D. and Bartram J. (2005). Water Safety Plans: Managing Drinking-Water Quality from Catchment to Consumer. WHO, Geneva 2005.

Dominguez-Chicas A. and Scrimshaw M. (2010). Hazard and risk assessment for indirect potable reuse schemes: An approach for use in developing water safety plans. *Water Research*, **44**, 6115–6123.

DWI (1999). Drinking water inspectorate: Information letter 10/99 – 25 June 1999. The Water Supply (Water Quality) (Amendment) Regulations 1999: Cryptosporidium in Water Supplies.

DWI (2004). Drinking water inspectorate: Information letter 06/2004 – 11th May 2004. Microbiological quality of water leaving treatment works and in service reservoirs: Drinking Water Safety Plans and Regulation.

Financial Reporting Council Ltd (2014). Guidance on Risk Management, Internal Control and Related Financial and Business Reporting, London, UK, 23 pp.

Gissurarson L. R. and Thoroddsson G. (2000). Hazard plan for disasters at reykjavik water works. *Water Supply*, **18**, 118–120.

Godfrey S. and Howard G. (2005). Water safety plans: Book 1 planning water safety management for urban piped water supplies in developing countries. Water, Engineering and Development Centre, Loughborough University, UK, 2005.

Jalba D., Cromar N. J., Pollard S. J. T., Charrois J. W., Bradshaw R. and Hrudey S. E. (2014). Effective drinking water collaborations are not accidental: Interagency relationships in the international water utility sector. *Sci. Tot. Environ.*, **470**, 934–944.

Hutter B. M. (2005). The attractions of risk-based regulation. Accounting for the emergence of risk ideas in regulation, ESRC Centre for analysis of risk and regulation, London School of Economics, London.

International Water Association (2004). The Bonn Charter for safe drinking water: Promoting best practice to manage drinking water quality effectively. IWA, Den Haag, The Netherlands.

Lindhe A. et al. (2011). Cost-effectiveness analysis of risk-reduction measures to reach water safety targets. *Water Research*, **45**, 241–253.

McKay J. (2003). Who owns Australia's water – Elements of an effective regulatory model. *Water Science and Technology*, **48**, 165–172.

NHMRC (National Health and Medical Research Council) (2004). Australian Drinking Water Guidelines (Online). Canberra: NHMRC. Available: http://www.nhmrc.gov.au/publications/synopses/eh19syn.htm.

NZMOH (New Zealand Ministry of Health) (2001). A Framework on How to Prepare and Develop Public Health Risk Management Plans for Drinking-Water Supplies (online). Wellington, New Zealand: Ministry of Health. Available: http://www.moh.govt.nz.

Pollard S. J. T., Gormley Á., Shaw H., Mauelshagen C., Hrudey S. E., Owen D., Miller G., Fesko P. and Pritchard R. (2013). Risk governance: An implementation guide for water utilities, (TC4363), Water Research Foundation, Denver, Co., US, 81 pp.

Renzetti S. and Dupont D. (2003). Ownership and performance of water utilities. *Greener Management International*, 9–19.

Rogers J. W. and Louis G. E. (2008). Risk and opportunity in upgrading the US drinking water infrastructure system. *Journal of Environmental Management*, **87**: 26–36.

Rouse M. (2007). Institutional Governance and Regulation of Water Services. IWA Publishing, London, UK.

Ruester S. and Zschille M. (2010). The impact of governance structure on firm performance: An application to the german water distribution sector. *Utilities Policy*, **18**, 154–162.

Saal D. S. and Parker D. (2000). The impact of privatization and regulation on the water and sewerage industry in england and wales: A translog cost function model. *Managerial and Decision Economics*, **21**, 253–268.

Saal D. S., Parker D. and Weyman-Jones T. (2007). Determining the contribution of technical change, efficiency change and scale change to productivity growth in the privatized english and welsh water and sewerage industry: 1985–2000. *Journal of Productivity Analysis*, **28**, 127–139.

Summerill C., Smith J., Webster J. and Pollard S. J. T. (2010). An international review of the challenges associated with securing 'buy-in' for water safety plans within providers of drinking water supplies. *J. Water & Health*, **8**, 387–398.

Summerill C., Pollard S. J. T. and Smith J. A. (2010). The role of organizational culture and leadership in water safety plan implementation for improved risk management. *Sci. Tot. Environ.*, **408**, 4319–4327.

Summerill C., Pollard S. J. T., Smith J. A., Breach B. and Williams T. (2011). Securing executive buy-in for preventative risk management – lessons from water safety plans. *Wat. Sci. Technol.: Water Supply*, **11**, 682–691

UK Water Industry Research (2002). Capital maintenance planning. A common framework. UK WIR Report Ref. No. 02/RG/05/, UKWIR, London.

Wallsten S. and Kosec K. (2008). The effects of ownership and benchmark competition: An empirical analysis of U.S. water systems. *International Journal of Industrial Organization*, **26**, 186–205.

WHO (World Health Organisation) (2004a). *Guidelines for Drinking Water Quality* (online). 3rd edn.

Unit 6

Corporate risk governance

6.1 OVERVIEW – PEOPLE MANAGE RISK IN ORGANISATIONS

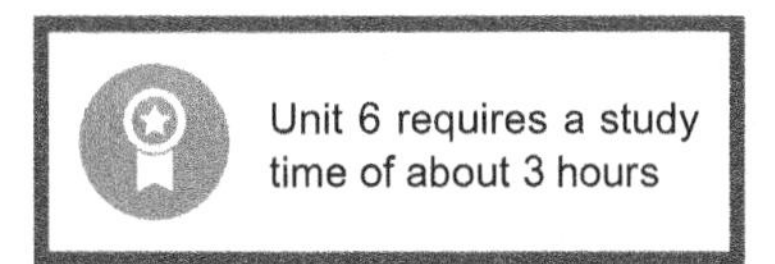

Managing risk well is a key competency for water utilities, and many utilities have established risk manager roles to coordinate their efforts. An essential requirement for utilities is to develop a preventative and anticipatory approach to managing risk and opportunity that ensures they are resilient to threats, whilst equally alive to business opportunities, such as those that emerge from technological innovation, for example. In practice, this means developing an organisational capability to connect operational activities to utility-wide, corporate (or enterprise) risk management programmes; to understand the impact of risk on a utility's corporate priorities; and then to forecast future risks into the mid- and long term so stakeholders can be confident in the utility's strategic plans designed to manage risk over the planning and investment cycles.

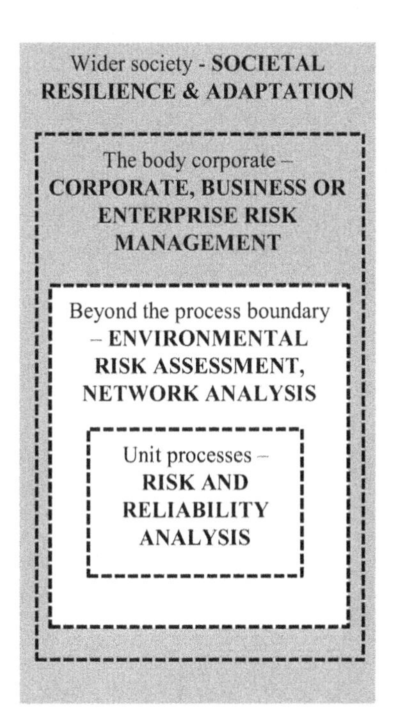

In this unit, we are concerned with the so-called 'body corporate' – the water utility organisation as a whole and the ability of its staff to analyse, manage, communicate and govern risks and opportunities. This level of risk management is variously termed 'enterprise risk management' (ERM), 'corporate risk management', 'strategic risk management' or 'business risk management'. These terms refer to the activity that organisations undertake to appraise their top-level risks that could pose a threat to their long-term corporate objectives. Given the organisational focus here, we adopt the term 'corporate risk governance'. This stresses the importance of the people and organisational frameworks that utilities, among other business sectors, use to ensure risks are managed throughout their organisational structures. Whilst the literature on corporate risk governance is replete with management terminology, business process diagrams and lists of strategic risks requiring management, there remain some fundamental practical principles that are core to good risk governance:

- People manage risk and opportunity, so clear lines of responsibility and an active sense of accountability among those individuals responsible for risk management are essential.
- Utility Boards and management executives set the tone of risk governance in their utilities, so the level of interest they take in how risks are managed is noticed, and often repeated, by others. Because they are somewhat at distance from operations, they can inadvertently become blind to issues deep in the business, and have to guard against this.
- As with all risk assessments, the origin and quality of the information used to inform corporate risk assessments is critical to the risk management

measures they inform. Auditable mechanisms for escalating and delegating risks are key to ensuring risk information is handled in a structured fashion.

- The organisational culture of the utility is important to the measured management of risk and opportunity. Some risks require active vigilance at all times; others might afford a more relaxed approach. A mature capability in risk governance can recognise the different characteristics of various corporate risks and design risk management measures suited to the risk appetite of the organisation.
- All organisations harbour latent flaws – precursors for risky situations that can become active when change happens. Rooting these flaws out through a continuous improvement programme, even when 'bad things' are not happening, is a responsible approach to good risk governance. Risk managers have to champion such activity and make convincing business cases for financing this type of preventative risk management.

6.2 A BASIC ASSUMPTION – THE PROTECTION OF PUBLIC HEALTH

Management scientists refer to the basic assumption of a business – the key defining purpose of its operation. Public health and environmental protection must be the primary goal of a water and wastewater utility. This goal may come under pressure from the practical need to manage a plethora of other business risks. Meeting higher service level expectations with limited resources, continually replacing assets and ageing infrastructures and managing public health threats alongside a growing portfolio of other business risks represents a demanding operating environment.

Consider the business climate of a large combined utility in England and Wales. Yorkshire Water Services (YWS) is responsible for the provision of water and wastewater services to 4.7 million UK customers. Over 95% of its customers are linked into the Yorkshire Grid asset network of resources, treatment and distribution pipes, all operationally controlled through YWS's production planning business process. This network of interconnected assets has over 1200 major elements including 147 reservoirs, 5 river sources, 80 boreholes, 86 water treatment works, 300 pumping stations, over 650 treated water storage reservoirs and 32,000 km of distribution pipelines to satisfy a demand of 1250 Ml/d. Maintaining these assets at a balanced level of serviceability to maintain supply security, water quality and excellence in product service is a central corporate business objective. Business risk management is central to securing service delivery.

EXAMPLE 6.1 MANAGING UTILITY INTERDEPENDENCIES

Operations in water utilities are dependent on services elsewhere, as the Cleveland (US) power cut demonstrated. A power outage on the 14th of August 2003 affected the North-Eastern States and parts of Canada. Cleveland Division of Water (CWD) saw the complete loss of power to their water treatment and distribution system. CWD serves about 1.5 million customers between 250 Mgal/d of water on a non-summer day and 300 Mgal/d on an average summer day. Cleveland CWD resumed normal operations 2 days. Following the incident, Cleveland CWD assessed the risk from supply chains in relation to their duties to provide drinking water and identified the critical need for uninterruptible power supply to their critical infrastructure.

EXAMPLE 6.2 IMPLEMENTING ENTERPRISE RISK MANAGEMENT

Louisville Water Company (LWC) supplies drinking water to >810,000 persons in Louisville and north-central Kentucky. The utility has 450 employees and is led by a team of executives and a Board of Water Works. LWCs in-house risk management approach has been influenced by the Committee of Sponsoring Organizations model (COSO) of the Treadway Commission which provided the assessment of "soft controls" to evaluate ethics and integrity within the company's culture, operating standards, employee competencies and empowerment.

Two teams manage the risk assessment process. The Business System Owners' Council assesses operational risk and the Executive Leadership Team has responsibility for strategic and financial risk. Risk ranking establishes the top risks requiring further analyses. Internal audit staff conduct risk-based audits in which the adequacy of risk control measures at LWC is evaluated.

Here then, our aim is to introduce and explore corporate risk management – the analysis and management of business risks faced by water and wastewater utilities. The discussion will take us beyond the operational aspects of the water business into aspects of business strategy and its implementation. Having worked through this unit, you should be able to:

- recognise and characterise strategic, programme-level (tactical) and operational risks;
- describe examples of each from the sector;
- understand the international drivers that are promoting good corporate governance and risk management; and
- summarise the key actions for utilities to ensure good governance of business risks.

6.3 THE PRACTICE OF RISK GOVERNANCE IN UTILITIES

Delivering reliability within a multi-stakeholder, institutional and business context, (Figure 6.1), in which expectations are rising is challenging. Privatisation, sector globalisation, increased competition, emerging technologies, increasingly stringent regulation and the trend towards financial self-sufficiency are transforming the water and wastewater sector, posing new risks and opportunities. The drivers for an improved governance of risk come not only from environmental and public health regulation, but also from the impacts of corporate misdemeanours that have demonstrated some organisations to be wanting with respect to their abilities to manage risk.

Over recent years, a series of risk management frameworks have emerged (Table 6.1), across the technical domains that a utility must consider. Driven largely by regulatory agencies in support of clearer environmental assessment or formal permitting requirements, these frameworks promote the use of tiered approaches to risk assessment; the use of options appraisal for selecting risk management measures (cost/benefit; social perceptions); and resource-efficient approaches to risk analysis that build on a fundamental understanding of the risk problem prior to the adoption of sophisticated techniques. One challenge is that a complex assembly of risk assessments across multiple classes of risk has grown up organically from this volume of practice, that now require synthesis and integration at some corporate level in order to make sense of the whole picture – a nontrivial task.

In response, many utilities in the sector are promoting an enterprise-wide approach to risk management. This requires:

(i) integrated frameworks for the management of internal risks, (e.g. from ageing infrastructure), and external, (e.g. from market processes or competitor actions), risks to the utility;
(ii) the support of Board level, executive management and operational staff, as well as that of external stakeholders; and
(iii) the effective communication of risk and engagement within decision-making processes both within companies and with external stakeholders.

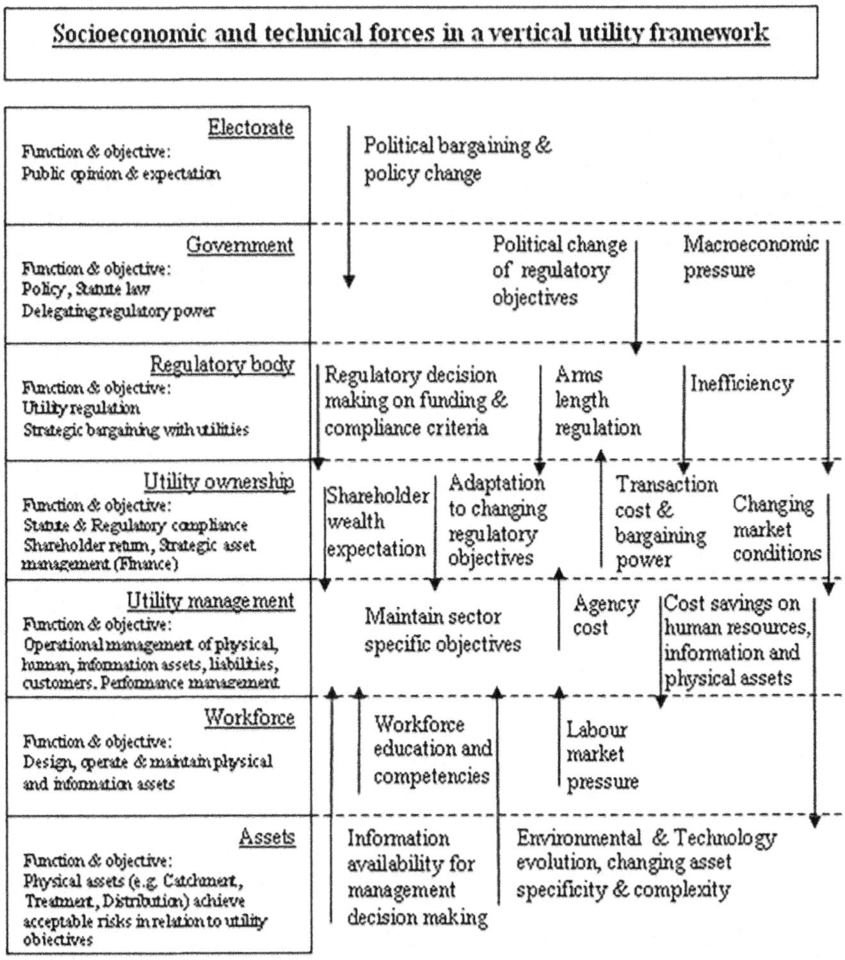

Figure 6.1 Utilities do not operate in isolation and their risks are intertwined with those of other stakeholders. This figure summarises just some of the socio-technical forces acting between utilities, the business world and wider civic society (after Woo & Vicente, 2003).

Managing corporate risk is as much about organisational change as it is about having structures in place, risk champions and tools and techniques for risk analysis that support better decision-making. Developing organisational cultures responsive, and not wholly averse, to risk is also a challenge. Critical features of the relationship between risk management, organisational performance and culture include:

- the importance of openness, transparency, engagement, proportionality, precaution, evidence and responsibility to good decision-making;
- the critical role of taking a long-term perspective in assessing the potential indirect consequences of management actions; and
- a widely held experience that 'hard' quantitative risk analysis tools used in isolation of transparent decision-making does little to gain public confidence and can result in the long-term erosion of trust.

The hierarchy that exists in organisational structures requires the effective management of risk at strategic, programme (tactical) and operational levels within the business (Figure 6.2). Typically, there are different accountabilities assigned for managing these risks, such that the chief financial officer/financial director and Board have overall corporate responsibility, supported by an internal audit or control function for the management of strategic risks; executive management for programme level risks (e.g. asset management, maintenance planning) and operational (e.g. site) managers for operational risks (e.g. plant performance).

Table 6.1 Risk management frameworks across different technical domains.

	Area of use	Framework	Reference
Risk assessment frameworks	Ecological risk	Framework for ecological risk assessment	USEPA (1992)
		Guidelines for ecological risk assessment	USEPA (1998)
	Microbiological risk	Conceptual framework for human health (chemical) risk assessment	NAS (1983) [US]
		Conceptual framework for assessing the risks of human disease following exposure to waterborne pathogens	ILSI (1996)
		Revised framework for microbial risk assessment	ILSI (2000)
	Public health risk	Risk assessment model	enHealth (2002)
Frameworks for risk-based decision-making/risk management	Public health risk	Framework for environmental health risk management	USPCRM (1997)
	General	Risk management process	AIRMIC *et al.* (2002)
		AS/NZS 4360:1999	AS/NZS (1999)
		Risk management overview	
		AS/NZS 4360:1999	
		Risk management process	
		AS/NZS 4360:1999	
		Risk treatment process	
		AS/NZS 4360:2004	AS/NZS (2004)
		Risk management process	
		CAN/CSA-Q850-97	CSA (1997)
		Q850 Risk management decision making	
		Benchmark risk management framework	NERAM (2003)
Regulatory prescribed frameworks for risk management	Water safety	Framework for the management of drinking water quality	NHMRC (2001, 2004)
		Public health risk management plan for drinking water supplies	NZMOH (2001)
		Multi-barrier approach	NERAM (2002)
		Water safety plan	UKWIR (2003)
		Guidelines for drinking water quality	WHO (2004)
	Capital maintenance planning	Common Framework for capital maintenance	UKWIR (2002)
	Security	Vulnerability assessment model	NIJ (2002)
Frameworks for corporate governance/enterprise risk management	Corporate governance	Implementing Turnbull	ICAEW (1999b)
	Enterprise risk management	'Stepping stones' along the enterprise-wide risk management journey	Deloach (2000)
		'Launch-pad' to enterprise-wide risk management	
		International Standards Organisation	ISO (2009)
		Enterprise risk management framework	Lam (2003)
		Enterprise risk management framework	COSO (2004)
		Integrated view of governance, ERM and compliance	Price Waterhouse Coopers (2004)
		Four stages of ERM process	Lam (2005)

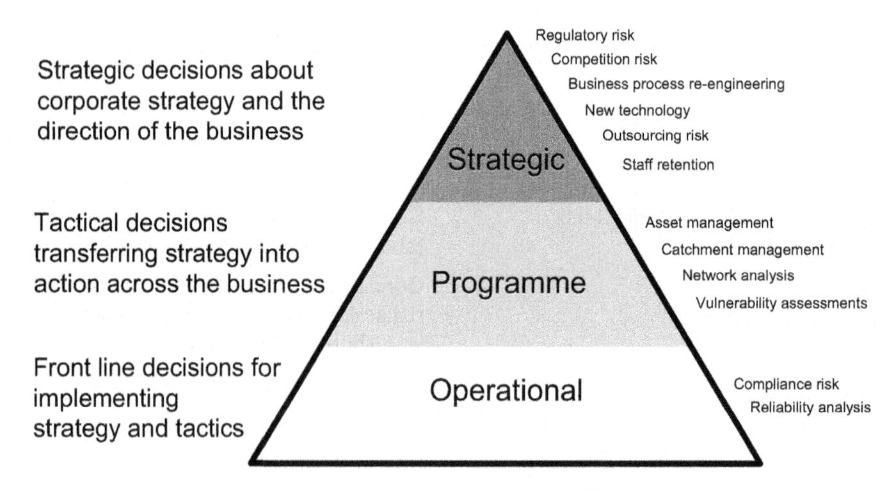

Figure 6.2 The risk hierarchy applied to the water utility sector.

We do not assume utility risks fit neatly into these simplified categories, or necessarily respect the boundaries of management hierarchies or organisational structures between them, but for ease of discussion we might compare risks at the strategic level with routine operational risks on the 'front line' of utility operations. Further, we recognise that operational risks can easily escalate to become strategic threats to a utility, so it would be naïve to assume risks 'stay where they belong' within Figure 6.2, or any other categorisation for that matter.

Risks at the strategic level might include major commercial and financial risks, such as those associated with infrastructure investment; merger and acquisition activity; company reputation; outsourcing; and the long-term viability of investment decisions.

At the programme or tactical level, we are concerned with techniques used to evaluate the risks posed by a similar hazard at a variety of locations (e.g. mains bursts, progressive failure of filter media – in asset management, for example) or with the wide variety of risks existing within a watershed. Tactical, programme-level risks are those that are company-wide. The availability of geographic information systems has facilitated the ease with which such assessments can be performed. Programme level risk assessments are concerned with the implementation of strategies across multiple sites and geographic regions (catchment/watershed planning). A typical tactical issue is the optimisation of workforce planning (e.g. asset inspection or maintenance scheduling): how should a limited resource (staff, time) be most effectively utilised in managing the risks that matter? Using risk-based resourcing, human (and financial) resources are targeted towards addressing higher risk activities. The presumption is that by moving to risk-based resourcing – here, by focusing on poorly performing assets the failure consequences from which are higher – yields greater risk reduction per unit resource input. Many water companies now have risk ranking procedures in place to inform maintenance scheduling and capital investment.

Credit: istockphoto © ismagilov

Operational risk analysis has been our concern to date in this text. Here we are concerned with the assessment of risks associated with specific operations at plant level – for example, the risk of failure of a device or process component, or the risk of exceeding a particular water quality standard. These risks need to be managed in the context of a business operation. Utility managers manage cash flow, whether raised through public or private sector means, by optimising expenditure and income in the context of water and wastewater services. Events inevitably occur that reduce income and cause increased expenditure. For example, failures in an infrastructure/distribution network causing leaks lead to increased expenditure on leak detection and pipe repair, refurbishment or asset replacement. In order to ensure the delivery of strategic, corporate objectives, risk management is required for at least one, and usually several of the following types of risk:

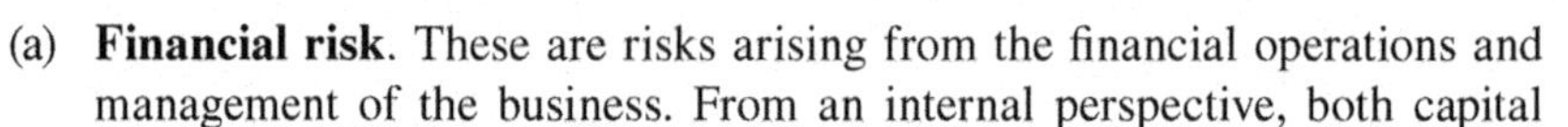

(a) **Financial risk**. These are risks arising from the financial operations and management of the business. From an internal perspective, both capital and operating costs are reduced by carrying out assessments during the design stage of a scheme, whilst operating and maintenance costs can be optimised on existing plant. External financial risks are derived from market processes, (*i.e.* currency rate fluctuations), and are of rapidly increasing significance to water utilities, given the increasing need for self-financing and the twin trends of privatisation and globalisation.

(b) **Commercial risk.** Formerly considered immune to such risks owing to their public sector monopoly, utilities are no longer insulated from competition or financial instability. They face an increasingly demanding public with powers to make significant changes if unsatisfied. A serious accident can depress the share price in publicly quoted companies, persuading Boards to make management changes.

(c) **Public health risks.** Failure or inadequacy of the treatment and distribution process can result in an interruption of supply or derogation in water or wastewater quality, (microbiological or chemical). The underlying causes may include source contamination, human error, mechanical failure or network intrusion. The consequences are often immediate – there is very little time to reduce exposure – and so the impacts can affect a large number of people simultaneously. The financial costs to the community of the fatal Walkerton outbreak, for example, were in excess of Cdn$65 million, with one time costs to Ontario estimated at more than Cdn$100 million. The loss of consumer confidence has been enormous.

(d) **Environmental risk.** Equipment failure or human error can lead to environmental impacts including discharges to the atmosphere, ground or the water environment. These may occur directly or as a result of actions to mitigate the results of the failure, such as discharges of polluted water from a tank. Similarly environmental change can affect water quality. The relationship between watershed management, source quality and water quality at the tap is inevitably one characterised by a need for sound risk management.

(e) **Reputation risk.** For most water utilities and regulators, the biggest fear is the loss of consumer confidence. If an organisation has target levels in its customer charter, considerable time and effort can be taken up by customer service functions explaining to customers and the media when incidents occur. The Sydney Water crisis of 1998 was estimated to have cost Sydney Water Corporation over $37 million in direct costs with contingency costs estimated at over $100 million for an episode with no health consequences.

(f) **Compliance/legal risk.** Legislation sets out minimum standards for water quality, the handling and storage of treatment chemicals, the discharge of wastes and the health and safety of the operational staff and the people living nearby. Aside from the inherent risk of failure to comply with legislation, (with all its associated consequences), there is the risk arising from uncertainty regarding the future actions of the regulator/legislator. Even when there is no legislation covering some aspects, there can be claims of negligence against operating companies. Litigation for civil damages were prominent features in both the Walkerton outbreak (settled out of court) and the Sydney Water Crisis (largely dismissed, but costs still incurred).

SUPPLY CHAIN RISK MANAGEMENT

The advent of 'outsourcing' in the utility sector has meant a need for managing risks in the supply chain. Such an assessment might consider the reliability of the supply chain, its vulnerability, the business recovery (down) time required to secure normal operation, the provision of capabilities in operational emergencies and the criticality of the operation to the overall business.

Assessing risk across these and other categories becomes essential for optimising operations and cost/benefit trade-offs. Many utilities now have risk management committees in place that monitor and report on priority risks, principally in response to requirements on internal financial control (audit). Figure 6.3 illustrates the internal structures in place (as of 2005) within United Utilities plc for risk governance. Beyond company boundaries, there are also important risk interfaces with other institutions including the regulators (licenses to operate), host communities (being a good corporate neighbour), and the capital markets, (raising finance).

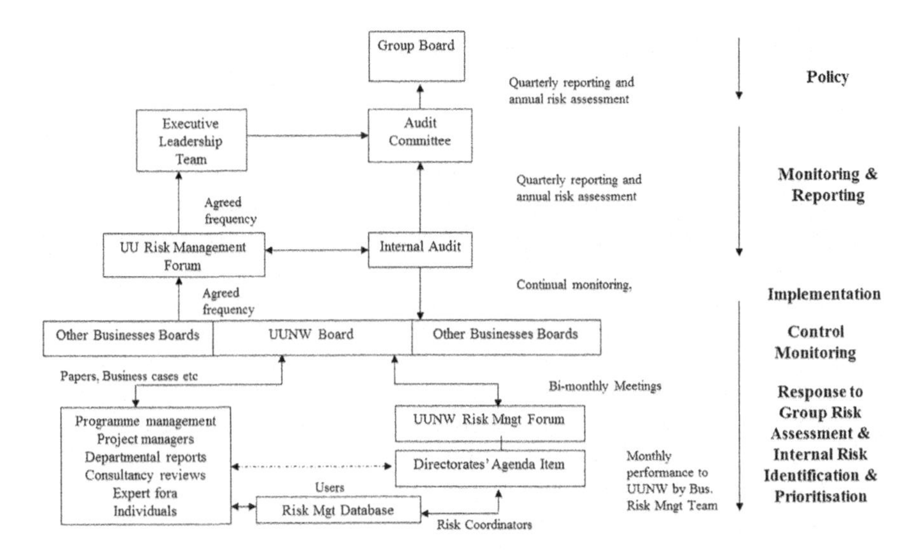

Figure 6.3 Risk governance arrangements at United Utilities plc (after Miller, 2005). This figure shows many features common to good corporate risk management, in this case for a privatised utility with a group structure. Note the presence of the risk management database to capture risk knowledge and inform the priorities of the regional (north west) risk management forum and then Board. Note formal review and escalation routes to the Executive Leadership Team, Group Board and parallel routes to Internal Audit. Note the structured 'rhythm' of reporting between the various organisational levels and (right hand side, in bold), the different locus of attention from risk policy, through risk management reporting, implementation to group risk assessment then risk identification deep in the business.

6.3.1 Assessing corporate risks in practice

In practice utilities must garnish the risks from their business units or functions and assemble them for risk analysis – in order to manage those who's significance is uncertain. There is controversy among practitioners and commentators of ERM as to the representativeness and true value of this practice; and whether the escalation and assessment of risks in this way, particularly at distance from the location where the risks may be realised, actually improves risk management. This said, most utilities now have in place an organisational structure similar to Figure 6.3 for appraising risks vertically, up through the business.

The conventional approach, though there are many variants, is to use local risk committees within the business to appraise risks locally, prioritise them and escalate high priority risks above a certain risk threshold for a higher-level analysis. Note there will be different probability and consequence scales for different risk categories in (a) to (f) above. Wide use has been made of risk matrices and so-called risk 'heat maps' that depict risks of increasing magnitude as Red (act now); Amber (analyse further, usually in the ALARP region) or Green (accept/monitor). At the corporate level, a register of risks is thus developed, across a set of categories, which represents key threats to the corporate objectives of the utility. These corporate risks can then be appraised further, actions developed for their management to acceptable levels of residual risk, and accountabilities assigned for the implementation of risk management measures, as deemed necessary. Periodic reappraisal (and occasional independent audit) ensures 'risk owners' have implemented, and checked the effectiveness of, the measures for which they are accountable.

Various stakeholders maintain an active interest in the capability of utilities to appraise and manage risks in this way and put in place the full set of organisational arrangements (Figure 6.4) to ensure the core risk assessment and management processes are functional. Investors seeks assurance that capital invested will provide a return and regulators seek confidence of a utilities anticipatory capability to preventatively manage public risk – that is risks to public health and the environment, as well as 'value-for-money' for the public purse. Customers and citizens wish to be assured that utilities are acting in their interests through the active management of risk.

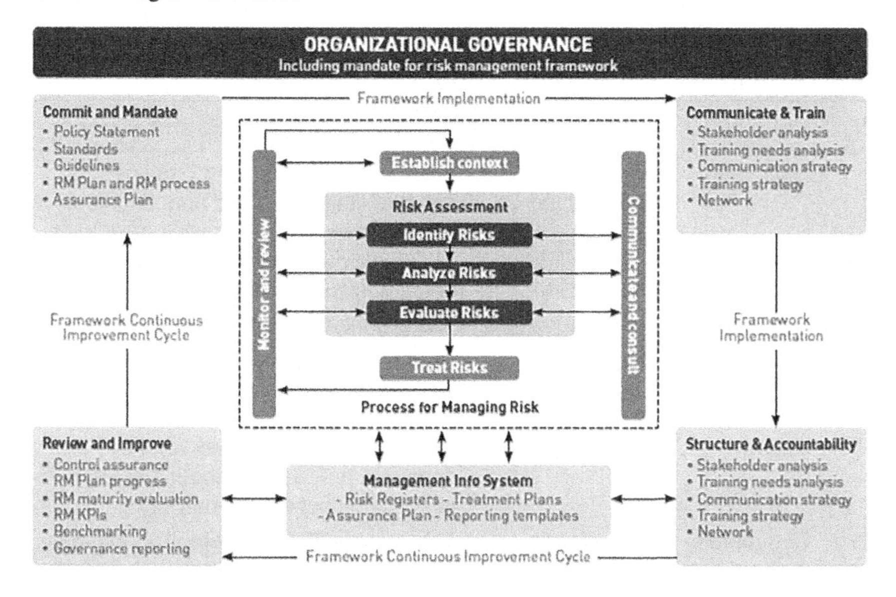

Figure 6.4 Corporate risk governance; the ability to not only identify, analyse and evaluate the significance of risk, but also to put in place the training, accountability, review and commitment to ensure risk and opportunity is actively managed.

6.4 DEVELOPING CAPABILITIES IN CORPORATE RISK MANAGEMENT

One theme throughout this text has been the reminder that risk assessments in themselves do not guarantee risk reduction. Left without their recommendations implemented, they represent an important analysis but may serve to increase legal liability if foreseeable and unaddressed failures subsequently occur. The follow-through of risk management actions through an organisation remains critical therefore and, because this requires line management tenacity, is a measure of organisational attentiveness to risk. Managing risk competently, wisely and by targeting the risk critical elements of a system for maximum risk reduction is what counts in practice. But how mature is the water and wastewater utility sector in assessing and managing risk and what makes for a sound organisational competency in risk management?

To understand organisational competencies in risk management, we must first look to the organisational *capability* a utility possesses in risk management. Because most water utilities manage risk by virtue of the routine provision of safe drinking water, we are generally concerned with the relative *maturity* of their capability, rather than just its presence or absence. We are concerned with an organisational ability to act wisely – to anticipate when things might go wrong and act in a preventative fashion to manage public health risks, for example. In practice, this means anticipating and responding to changes in parameters that might induce a loss of performance – reduced staffing levels, extreme rainfall events, shrink-swell cycles in ground conditions – the combination of root causes that impact on the performance of water and wastewater services.

One way to evaluate organisational capability is to develop a benchmark against which organisations can be measured. 'Capability maturity models' codify industry practice so distinctions can be made between organisations and improvements identified and made. They can also be used to assess the degree of competency with which an organisation performs the key processes required to deliver a business function such as risk management (Table 6.2). This is then represented by a level

CAPABILITY MATURITY MODELS

Capability maturity models are simplified representations of organisational disciplines (e.g. software design and engineering) that codify industry practice within a maturity framework. They allow distinctions between organisational capabilities (e.g. the ability to manage risk) by reference to the maturity of the processes in place. Those models developed for risk management maturity typically examine a small number of essential risk management processes within an organisation at different levels of maturity. Visually, results are presented as 'radar plots' that illustrate the overall pattern of maturity across the management processes.

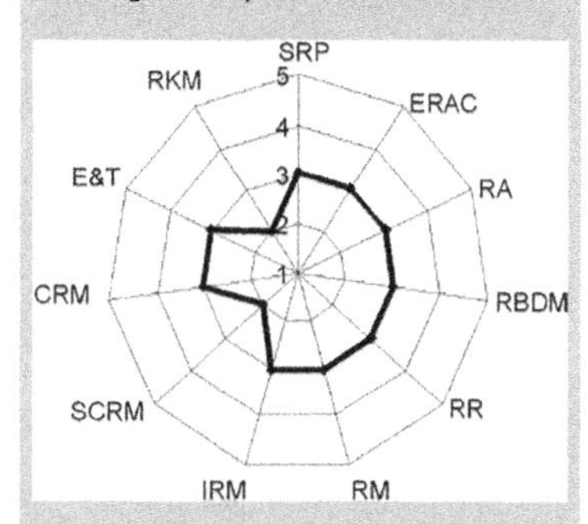

	Processes
Core	Strategic risk planning (SRP)
	Establishing risk acceptance criteria (ERAC)
	Risk analysis (RA)
	Risk based decision making and review (RBDM)
	Risk response (RR)
	Risk monitoring (RM)
	Integrating risk management (IRM)
Supporting	Supply chain risk management (SCRM)
	Change risk management (CRM)
Long-term	Education and training in risk management (E&T)
	Risk knowledge management (RKM)

of maturity. Level 5 (high) organisations exhibit 'best practice' in discharging the chose function. They are capable of learning and adapting and they use their experiences to correct problems and change the way they operate. In contrast, Level 1 (low) organisations are learner organisations with non-standard and largely uncontrolled processes.

Table 6.2 Interpretation of maturity levels.

N	Maturity	Mode/style	Process Characteristic and Effect
5	Optimised	Adaptive, double loop learning	The organisation is 'best practice', capable of learning and adapting itself. It not only uses experience to correct any problems, but also to change the nature of the way it operates
4	Managed	Quantified, single loop learning	The organisation can control what it does in the way of processes. It lays down requirements and ensures that these are met through feedback.
3	Define	Measured, open loop	The organisation can say what it does and how it goes about it but not necessarily act on its analyses
2	Repeatable	Prescriptive	The organisation can repeat what it has done before, but not necessarily define what it does
1	Ad hoc	re-active	Characterises a learner organisation with complete processes that are not standardised and are largely uncontrolled
0	Incomplete	Violation	Incomplete processes, criminal or deliberate violation tendencies

RISKS AND OPPORTUNITIES OF WATER PRIVATISATION

Since privatization in England and Wales the water companies have rationalized their workforce. Increasing pressures to minimize cost has meant reduced staffing levels on operational sites and increased remote operation. The privatised water utilities face:

- regulatory pressure to operate more efficiently and promote the reduction of operational cost, e.g. releasing labour
- a need to optimise operational cost, plant reliability, customer service and asset value
- increasing the workload of operational staff with the potential consequence of reducing plant maintenance
- migration of the work force into other industries (problems of staff retention)
- loosing operational expert knowledge as in-house experts retire or take severance
- higher competition in the employment market with impact on labour cost.

Management of these risks has provided opportunities to:

- reduce staffing to an acceptable level of plant reliability
- provide optimal levels for maintenance to sustain the asset base at maximum returns over the asset life
- balance the long term cost of asset performance, condition and replacement with the costs for staff to operate and maintain plant and equipment
- optimise the substitution of labour with technology.

Many utilities seek to improve the processes involved with managing their risks as an effective means of reducing exposure and improving risk-based decision making within their organisations. Understanding one's own risk management maturity has value in that it may:

(i) assist organisations in formalising their appetite for risk;
(ii) help formalise and make more explicit the role of the group risk manager;
(iii) provide the opportunity for a 'climate' check on the implementation of risk management procedures on the ground within the organisation;
(iv) provide evidence to support corporate level statements on risk.

6.5 DEVELOPING A RISK MANAGEMENT CULTURE

Much has been written on safety culture in the light of organisation accidents since the 1970s, much of it coincident with good risk management practice. Developing a risk management culture that is sustaining and continues to learn and improve in face of the inevitable peaks and troughs of organisation performance requires:

- leadership;
- procedures;
- an appetite for conservative decision-making where safety is put first even under pressure;
- a culture of sharing reported 'close calls';
- good communication at the appropriate level;
- an open, learning organisational culture able to benchmark itself against the best-in-class;
- systematic competency checking;
- effective management of organisational change; and
- the ability to prioritise.

Most incidents have deeply-rooted causes that are a combination of technical failures, an incapacity to manage change and the underlying values within, or market forces acting on, an organisation; as catalogued for waterborne outbreaks by Hrudey and Hrudey (2004). Adverse events, when they occur, represent an organisational failure to convert hindsight into foresight – that is, to learn from pat events and act upon this learning. Incidents typically occur:

- when there is a loss of institutional foresight and corporate memory;
- in the face of strong market pressures for efficiency gains;
- when there are considerable elements of outsourcing;
- where organisations fail to maintain their status as an 'intelligent customer';
- with loss of internal technical expertise; and particularly
- during, or following periods of business re-engineering.

Cost pressures, priority-based working and changes that are rushed are all circumstances that can generate accidents. For organisations to become resilient and mindful of these factors, vigilant and aware that is, they must anticipate and circumvent threats to corporate objectives and manage pressures and conflicts between corporate performance and the risks that threaten it. Modern management culture sets a strong impetus on doing more for less ('lean') and on maintaining business continuity. Middle managers may find it difficult to challenge this philosophy in managing risk issues up to the executive management or Board. When they do, risks that not quantified in monetary terms may inadvertently receive restricted attention at Board meetings and so lie dormant, unaddressed within the organisation as latent causal factors for an incident sometime in the future.

This chapter is about corporate risk governance and we stress the importance of implementing risk management measures deep in the business. These measures, acting as barriers to risks being realised in a utility, are usually manifold and, just like the multi-barrier approach to water treatment discussed earlier, are themselves susceptible unless maintained by people within organisational structures. Without 'preventative maintenance', they risk developing faults that allows hazards to be realised. So we start to view organisational risk management, at least for those events that we seek to guard against, as set of corporate barriers in place to prevent adverse outcomes.

The 'Swiss cheese' model of accident causation depicted in Figure 6.5 is a powerful visual model in which the layers of cheese represent barriers or defence stages to prevent failures; and the holes, (hence the Swiss Cheese label), represent deficiencies in those barriers. Reason (1997) draws a distinction in this model between *active failures* and *latent conditions*, an important distinction that we retain. Applying this model to a water quality incident, the holes are not static, but are changeable in response to a variety of factors such as operator actions, treatment performance variations and external conditions (weather, water quality variation).

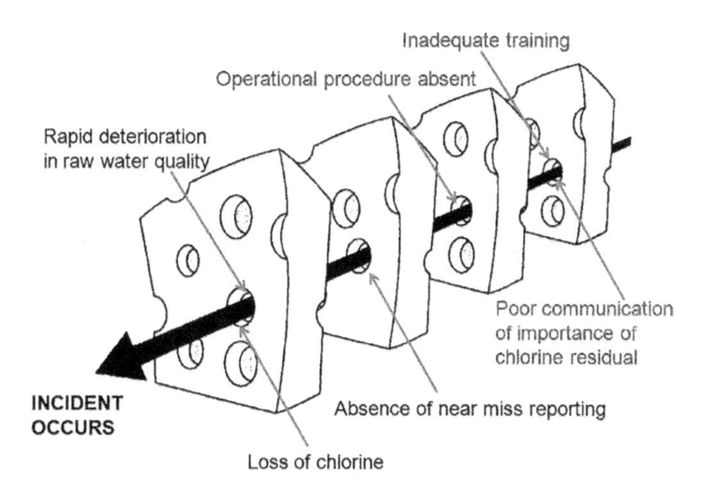

Figure 6.5 Swiss cheese model of organisational accident, applied to the Walkerton tragedy (after Reason, 2000; Hrudey *et al.*, 2006).

6.5.1 Example – Lessons from Walkerton

This Swiss Cheese model is well suited to understanding the array of factors that ultimately contributed to a disaster. Consider, for example, the mismanagement of the public water supply in Walkerton, Ontario (O'Connor, 2002a,b) that caused 7 deaths, 27 cases of haemolytic uremic syndrome mainly among children and over 2300 cases of gastrointestinal illness in May 2000.

In Walkerton, Ontario, the shallow (5.5 – 7.4 m) well that became contaminated was drilled 22 years before the outbreak, yet it was recognized by the hydrogeologist who commissioned it in 1978 to be at risk of agricultural contamination. He recommended acquiring a buffer zone, which was never done, and stressed the need for chlorination because the commissioning pump test had produced faecal coliforms within 24 hours. This well was approved for operation, eventually with a condition of maintaining a chlorine residual of 0.5 mg/L for 15 minutes, with chlorine residual to be measured once a day. When heavy spring rains washed manure into the shallow aquifer, the well that was being chlorine dosed at less than 0.5 mg/L was overwhelmed with an organic matter loading and the marginal disinfection completely negated. The opportunity to detect this disaster, by monitoring the chlorine residual and recognizing the absence of a residual, was lost because the operators had no understanding of the risks facing this drinking water supply or of the importance of chlorination to managing those risks. Their lack of knowledge was a factor in their failure to perform the required monitoring, thereby allowing Walkerton residents to consume water for at least nine days longer than necessary before a "boil water advisory" was issued.

A summary of some of the most obvious active failures and latent conditions that contributed to the Walkerton disaster is presented in Figure 6.4. These illustrate a range of contributing interactive factors that ought to be considered by any water utility that is seeking to minimize its risk of catastrophic failure. If the drinking water industry wishes to learn from this and similar tragedies (Hrudey & Hrudey, 2004), it needs to routinely document failures and near failure incidents to analyse them for the active flaws and latent conditions that contributed to the development of these incidents. In this way, a responsible programme of reducing active flaws and controlling latent conditions can become the foundation of a preventive risk management approach.

EXAMPLE 6.3 MANAGING RISK PROACTIVELY – WATER DEMAND UNDER FUTURE CLIMATE SCENARIOS

Assessing the vulnerability of water resource yields to climate change involved the application of regional factors for changes in yield – from groundwater recharge and stream flow, to observed recharge and flow data. The UK was divided into six regions and for each region, flows simulated in a number of catchments under current, and a range of possible, future climate conditions.

The result was a set of perturbed flow series, indicative of conditions under the changed climate of the 2020s. The potential reduction in late summer/autumn runoff was clear in S England, as was the possible increase in flows throughout the year in N England, Scotland and N Ireland. This approach provided water managers in the UK with a rapid screening of the potential risks from climate change on water resources in the face of significant uncertainty.

It is clear from events like the Walkerton tragedy, that leadership and management are key to establishing a risk management culture, in terms of the expectations and example that are set, or not. But how do organisations proactively develop a risk management culture without having first to suffer a major accident? How do we force ourselves to ensure risk issues are treated seriously? And how can we usefully process the volumes of risk information gathered by risk assessors and managers to make sense of it for accident/incident prevention? When should executive managers listen to the challenge from below? Above what threshold should they then act? And are risk managers arguably an additional source of risk because, in taking institutional responsibility for coordinating risk management, they risk absolving others of their individual responsibilities for risk management?

These are critical organisational questions germane to the organisation practice of risk management and requiring management research. They remind us that managing risk extends beyond the existence of an analytical capability or the effective implementation of procedures. It requires wisdom, time for reflection, the active management of risk knowledge and corporate mindfulness so that preventive approaches to risk management can be promoted that are creative and forward-looking. Weick and Sutcliffe (2001) characterise 'mindfulness' in organisations that are: (i) preoccupied with failure and the root causes of it; (ii) reluctant to (over)simplify; (iii) sensitive to operations; (iv) committed to resilience; and (v) deferential to expertise over management hierarchy. For water utilities seeking to develop mindfulness:

- informed vigilance must be actively promoted and rewarded;
- an understanding of the entire system, its challenges and limitations needs to be promoted and actively maintained;
- effective, real-time treatment process control, based on understanding critical capabilities and limitations of the technology, must become the basic operating approach;
- fail-safe multi-barriers need to be actively identified and maintained at a level appropriate to the challenges facing the system;
- close calls should be documented and used to train staff about how the system responds under stress and to identify what measures are needed to make such close calls less likely in future;
- operators, supervisors, lab personnel and management should all understand that they are entrusted with protecting the public's health and should be committed to honouring that responsibility above all else;
- operational personnel need to be afforded the status, training and remuneration commensurate with their responsibilities as guardians of the public's health;
- response capability and communication need to be improved, particularly as post 9–11 bioterrorism concerns are being addressed; and
- an overall continuous improvement, total quality management (TQM) mentality should pervade the organisation.

The benefits of this approach are long term and can be difficult to quantify, but we can look to other sectors such as the offshore oil and gas sector (Figure 6.6) where demonstrable reductions in business interruption losses have been secured through progression to a risk management culture.

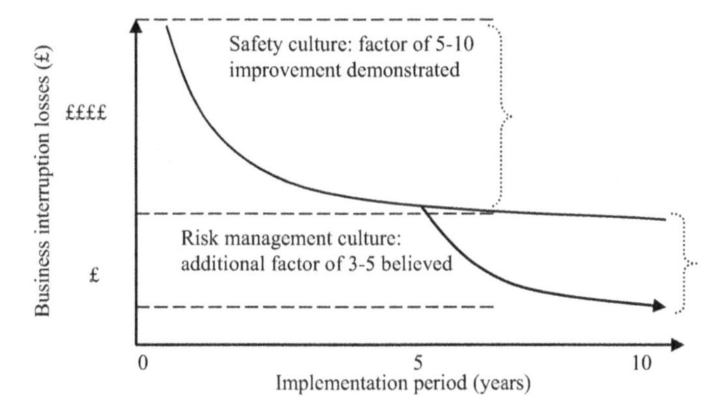

Figure 6.6 Scale of business interruption losses potentially achievable through transition to a sound risk management culture.

In this chapter, we develop a richer understanding of the relationship between business risk, organisational culture and the management of risk knowledge from within the organisation for the purposes of improved, risk-informed decision-making (Figure 6.7).

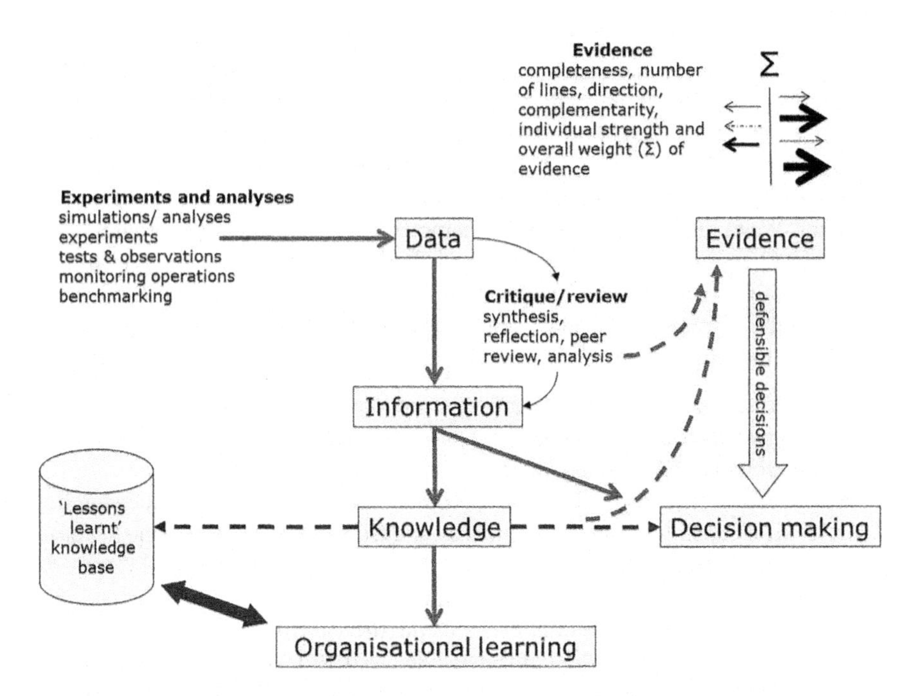

Figure 6.7 Making better decisions – the relationship between risk analysis, risk knowledge and evidence-based decision-making.

Building a mature capability in risk management rests on an ability to 'bank' and 'recall' organisational learning from the lessons an organisation learns. These are founded on its corporate knowledge of effective decisions that it has made in the past and, by default, on its ability to draw on evidence for sound decisions making which, in turn, is supported by data and information of sound provenance. The evidence one uses from risk assessments for the purposes of sound decision-making is rarely uniform or unequivocal in real life. Risk analysts must be able to weigh distinct lines of evidence (Σ, Figure 6.6) of different strength and resonance in order to make well-informed evaluations of risk significance and therefore conduct either measured risk-taking or anticipate harm and thereby enact preventative risk management.

It follows that the manner in which utilities capture and assesses their risk knowledge, who does this and the escalation of their findings through the organisation are critical. Retained corporate expertise, well-functioning line management arrangements and a receptive ear to risks from the business become essential to the capture and retention of lessons learnt.

6.6 SUMMARY AND SELF-ASSESSMENT QUESTIONS

We have considered the broader perspective of corporate risk management, the institutional structures for managing risks and the models that explain how organisational accidents can occur. Finally, we have reviewed the requirements for mindful water and wastewater utilities. In unit 8, we extend this further to consider better decision-making as a whole. However, we first consider the management of opportunity and reputational risk in water utilities.

SAQ 6.1 Propose a structure for the governance of risk within a water utility. Suggest the regularity of reappraising risk assessments for each part of the business.

SAQ 6.2 Think about your own appetite for risk. How does this impact on you seeking new opportunities? Now consider what parallels there are for utilities seeking to (i) adopt a new technology e.g. membrane treatment; (ii) acquire a new water company in Asia. What due diligence might you adopt before making these decisions?

SAQ 6.3 List 5 risks the water utility sector will need to actively manage considering the demographic changes in the workforce population. How might the sector manage these risks?

6.7 FURTHER READING

Andersen T. J. and Roggi O. (2014). Value creation through risk management. Chapter 4 in: Managing Risk and Opportunity: The Governance of Strategic Risk-Taking (eds), Andersen, Garevy and Roggi, Oxford University Press, Oxford, UK, 187 pp.

Arena M., Arnaboldi M. and Azzone G. (2011). Is enterprise risk management real? *J. Risk Research*, **14**, 779–797.

Arnell N. W., Reynard N. S., King R., Prudhomme C. and Branson J. (1997). Effects of climate change on river flows and groundwater recharge – guidelines for resource assessment, report to UK Water Industry Research/Environment Agency; Environment Agency: Bristol, UK.

Black and Veatch (2013). 2013 Strategic Directions in the U.S. Water Industry, Black and Veatch Corporation, USA, 79 pp.

Cass Business School (2011). Roads to Ruin. A Study of Major Risk Events: Their Origins, Impact and Implications, Cass Business School, Airmic, London, 179 pp.

CDP (2014). From Water Risk to Value Creation. CDP Global Water Report 2014, 59 pp.

Committee of Sponsoring Organizations of the Treadway Commission (2004). Enterprise Risk Management Integrated Framework, Executive Summary, available at: http://www.coso.org/documents/coso_erm_executivesummary.pdf.

Cranfield University (2014). Roads to Resilience. Building Dynamic Approaches to Risk to Achieve Future Success, Executive Summary, Cranfield School of Management, Airmic, London, 12 pp.

Fraser J. R. S. and Simkins B. (2007). Ten common misconceptions about enterprise risk management. *J. Corporate Finance*, **19**, 75–82.

Hillson D. (ed.) (2006). The Risk Management Universe – A Guided Tour. British Standards Institution, London.

Hrudey S. E. and Walker R. (2005). Walkerton – 5 years later. Tragedy could have been prevented. *Opflow*, **31**(6), 1, 4–7 (June 2005).

Hrudey S. E. and Hrudey E. J. (2004). Safe Drinking Water – Lessons From Recent Outbreaks in Affluent Nations. IWA Publishing, London, 514 pp.

Hrudey S. E., Hrudey E. and Pollard S. J. T. (2006). Risk management for assuring safe drinking water Environ. *Intl.*, **32**, 948–957.

Hrudey S. E., Huck P. M., Payment P., Gillham and Hrudey E. J. (2002). Walkerton: Lessons learned in comparison with waterborne outbreaks in the developed world. *J. Environ. Eng. Sci.*, **1**, 397–407.

Jalba D. I., Cromar N. J., Pollard S. J. T., Charrois J. W., Bradshaw R. A. and Hrudey S. E. (2010). Safe drinking water: Critical components of effective interagency relationships. *Environ. Intl.*, **36**, 51–59.

Institute of Development Studies (2010). The Resilience Renaissance? Unpacking of Resilience for Tackling Climate Change and Disasters. A. Bahadur, M. Ibrahim and T. Tanner (eds), Strengthening Climate Resilience Discussion Paper 1, Brighton, UK, 45 pp.

IWA (2004). The Bonn Charter for Safe Drinking Water. International Water Association, London http://www.iwahq.org.uk/pdf/Bon_Charter_Document.pdf.

International Standards Organisation (2009). ISO 31000:2009, Risk management – Principles and guidelines. ISO, Geneva, Switzerland.

Luís A., Lickorish F. and Pollard S. (2015). Assessing interdependent operational, tactical and strategic risks for improved utility master plans. *Water Research*, **74**, 213–226.

Luís A., Pollard S. and Lickorish F. (2016). Evolution of strategic risks under future scenarios for improved utility master plans. *Water Research*, 88, 719–727.

MacGillivray B. H. and Pollard S. J. T. (2008). What can water utilities do to improve risk management within their business functions? An improved tool and application of process benchmarking. *Environ. Intl.*, **34**, 1120–1131.

Marsh and McLennan Companies (2013). 2013 Water Industry Insurance and Risk Benchmarking Report, Marsh Ltd, UK, 18 pp.

Mauelshagen C., Rocks S., Pollard S. and Denyer D. (2011). Risk management pervasiveness and organisational maturity: A critical review. *Intl. J. Business Continuity & Risk Manage*, **2**(4), 305–323.

Miller A. G. (2005). In S. J. T. Pollard, S. E. Hrudey, L. Reekie and P. D. Hamilton (eds) Proc. AwwaRF International Workshop "Risk analysis strategies for better and more credible decision-making", Banff Centre, 6–8th April, 2005, AwwaRF and Cranfield University.

National Association of Corporate Directors (2009). Report of the NACD Blue Ribbon Commission. Risk governance: Balancing risk and reward. Executive summary, NACD, 20 pp.

NHMRC (2004). Australian Drinking Water Guidelines. National Health and Medical Research Council. Canberra, ACT. http://www.nhmrc.gov.au/publications/synopses/eh19syn.htm.

O'Connor D. R. (2002a). Report of the Walkerton inquiry. Part 1. The events of May 2000 and Related Issues. Toronto, The Walkerton Inquiry: 504 pp. http://www.attorneygeneral.jus.gov.on.ca/english/about/pubs/walkerton

O'Connor D. R. (2002b). Report of the Walkerton inquiry. Part 2. A strategy for safe water. Toronto, The Walkerton Inquiry: 582 pp. http://www.attorneygeneral.jus.gov.on.ca/english/about/pubs/walkerton.

Pollard S. J. T., Strutt J. E., MacGillivray B. H., Sharp J. V., Hrudey S. E. and Hamilton P. D. (2006). Risk management capabilities – towards mindfulness for the international water utility sector. In: Water Contamination Emergencies: Enhancing Our Response. K. C. Thompson and J. Gray (eds), Royal Society of Chemistry Publishing, Cambridge. pp. 70–80.

Pollard S., Hrudey S. E. et al. (2007). Risk analysis strategies for credible and defensible utility decisions, Awwa Research Foundation research report 91168, Awwa Research Foundation, American Water Works Association and IWA Publishing, Denver, CO, US, 88 pp.

Pollard S. J. T., Bradshaw R. et al. (2009). Developing A Risk Management Culture – 'Mindfulness' in the International Water Utility Sector (TC3184). Water Research Foundation, Denver, CO, US, 117 pp.

Pollard S. J. T., Gormley Á., Shaw H., Mauelshagen C., Hrudey S. E., Owen D., Miller G., Fesko P. and Pritchard R. (2013). Risk Governance: An Implementation Guide for Water Utilities, (TC4363). Water Research Foundation, Denver, CO, US, 81 pp.

Reason J. (1997). Managing the Risks of Organizational Accidents. Ashgate, Aldershot, England, 252 pp.

Reason J. (2000). Human Error: Models and Management. *BMJ*, **320**, 768–770.

Rittenberg L. and Martens F. (2012). Understanding and Communicating Risk Appetite. Committee of Sponsoring Organizations of the Treadway Commission (COSO).

Sinclair M. and Rizak S. (2004). Drinking water quality management: The Australian Framework. *J. Toxicol. Environ. Health, Part A.*, **67**, 1567–1579.

Smyth M. (2004). A risky business. *Water & Environment Manager*, **9**(1), 16, December 2003/January 2004.

Taylor R. (2005). Presented at: 'Achieving a good safety culture – the people dimension in health, safety and environmental performance', Hazard Forum open meeting, 10th March 2005, London.

Weick K. E., Sutcliffe K. M. and Obstfelt D. (1999). Organizing for high reliability: Processes of collective mindfulness in: Research in Organizational Behaviour. Jai Press, Stanford, pp. 81–123.

Weick K. E. and Sutcliffe K. M. (2001). Managing the Unexpected – Assuring High Performance in An Age of Complexity. University of Michigan Business School, Josey-Bass Publ., San Francisco.

Westerhoff G. P., Pomerance H. and Sklar D. (2005). Envisioning the future water utility. *J AWWA*, **97**(11), 67–74.

WHO (2004). Water Safety Plans. Chapter 4. WHO Guidelines for drinking water quality 3rd edn. Geneva, World Health Organization: 54–88. http://www.who.int/water_sanitation_health/dwq/gdwq3/en/.

Wilby R. and Vaughan K. (2010). Hallmarks of organisations that are adapting to climate change. *Water Environment J.*, **25**, 271–281.

Unit 7

Managing opportunities and reputations

SOUND RISK MANAGEMENT AS A BUSINESS OPPORTUNITY TO CREATE VALUE

Risk management provides substantive opportunities to utilities – to increase shareholder value, optimise asset life and performance, improve regulatory relations and attract capital investment. Demonstrable risk governance is now an expectation of most utility stakeholders.

Unit 7 requires a study time of about 3 hours

7.1 INTRODUCTION

So far in this text, our discussion has been principally on the management of adverse risks; those that present negative impacts on a utility business. The modern debate on risk has been one dominated by managing the threats of unwanted, adverse consequences on businesses; e.g. adverse public health impacts, environmental damage and vulnerability assessments for water utility infrastructures. This aligns with a business risk paradigm adopted in the late 1990s (Figure 7.1) that focused on threats to businesses.

Figure 7.1 Typical 1990's business risk paradigm.

The history of risk, of human endeavour and of business however, is a story of measured risk-taking, of innovation and of seizing opportunities for reputational and competitive advantage. Organisations want staff to be creative, to seek out innovations and, when appropriate, to take measured risks rather than be constrained by an overly cautious approach. This is fine for the manufacturing sector, perhaps, or for new technology development, but in the business of public health protection, the balance between risk and opportunity management is more delicate. There is a convention of exercising precaution and conservative (risk averse) decision-making that has generally served the water and wastewater sector well and become ingrained in most utilities. However, there are also financial pressures that continually act to challenge conventional wisdom. Mature risk management in utilities must allow for the adoption of preventative risk management for certain risks alongside measured risk taking for others. Water and wastewater utilities need to make well-reasoned arguments for the installation of new technologies as well as maintain a risk aversion towards the accidental introduction of pathogens to the public supply. How then do we strike the balance between:

- ensuring sound and preventative risk management;
- remaining open to innovation and the opportunities it provides;
- operating in a resource-constrained environment; and
- ensuring staff are not so bound by risk management 'procedure' that they fail to think meaningfully about the consequences of their actions?

Part of the answer is about the risk management culture that organisations put in place and manage. We discuss this further in unit 8. Equally important is the corporate approach to innovation. Linked to risk and opportunity is the reputational impact that corporate decisions have on internal staff and external stakeholders. Because utilities work closely in the public eye, their decisions and their perceived management of public health and safety are under substantive and regular scrutiny. For public utilities, perceptions may be affected by the perceived intentions of elected officials; for private companies, by company profits made from managing a public 'good', (water), and the personal bonuses of directors. Managing the risk of reputational damage, or the opportunity for reputational enhancement therefore, is a growing feature of utility management.

MANAGING RISK AND OPPORTUNITY WELL CREATES LONG-TERM VALUE FOR ORGANISATIONS

The good governance of business risk protects operators, customers and other stakeholders from harm, builds reputational value and can generate competitive advantage for best in class organisations. This chapter is concerned with realising this value – recognising it, securing it and using it as a preparation for future business challenges. For most utilities that already have risk management frameworks and procedures in place, the strategic value of good risk governance will secured by aligning risk management with other business functions to inform and 'future proof' long-term corporate priorities.

Having the organisational maturity to handle various risks with distinct appetites for each infers clarity over which risks can be tolerated and which cannot. Risk appetite defines how much risk an organisation seeks to take, and is prepared to accept, in pursuit of its corporate objectives. The benefits attributed to defining a formal risk appetite statement are improved business performance, better stakeholder confidence and a more effective corporate risk function. But clarity over how risks are handled in the business is also essential to having empowered staff prepared to take local decision with confidence. Without a defined risk appetite, decision makers must use their judgement to determine if a risk is tolerable or not. Given such judgements are affected by individual, social, cultural and other factors, this approach may not provide a defensible basis for informed risk-taking and is unlikely to be aligned with corporate strategy. Having a clear risk appetite statement can ensure consistent criteria are deployed and, with effective monitoring, this can increase the prospect of unacceptable risks being escalated for mitigation.

How utilities deal with emergency situations is equally critical. Emergencies can offer opportunities to build or lose stakeholder trust. If response procedures are developed, employees trained, equipment obtained and practice exercises carried out, both management and staff will have the confidence that a credible, customer caring response will be made to a major incident. When this occurs, effective emergency management is a real opportunity to build and enhance the corporate reputation.

7.2 INCORPORATING OPPORTUNITIES

21st century definitions of risk allow also for the identification and management of upside risks – that is, the positive benefits (consequences) that arise from taking measured business risks. Opportunities are simply business risks with a probability and a positive business impact. Techniques for identifying business opportunities have been developed by management practitioners (see Pollard & Carroll, 2001). Many are qualitative but can still be developed to capture the power of distinguishing between the likelihood and positive consequences of business decisions.

Tools and techniques for opportunity management are increasingly applied for technological innovation and, as such, have application for water and wastewater utilities in the technology 'road maps' being created for the sector as it prepares for a low carbon, chemical free future. The consequences of business opportunity can be estimated in financial terms (likely savings, returns on investment, net present values etc.) and the probability of securing these opportunities are usually managed by progressing through a series of 'gated' decision points – stage gates – where there is a chance to reflect on the likely success of a project or opportunity as it progresses by reviewing the available evidence gathered to date. This approach has been in wide use within research, technology and development for many years.

There is also a broader dimension to viewing good risk governance itself as an opportunity to be exploited for reputational and competitive advantage. Utilities that manage risk well build reputational credit, regulatory support and citizen confidence. Often, these are secured by organisations that have recognised the corporate value of good risk management. Andersen and Roggi (2014) discuss several pre-requisites for value creation through risk management, including through:

- a willingness to invest resources in opportunity development under changing conditions;
- active organizational learning from key events, incidents, or situations, good or bad;
- a commitment to some element of strategic foresight, or an ability to anticipate change and unexpected events, the pre-conditions for which may build up over time;
- an agility to respond creatively to emerging risks and opportunities and identify viable solutions to these in real time, along with identifying adaptive paths for the organization going forward;
- an ability to make informed decisions on risk and opportunity that resource the right activities at various levels throughout the organisation;
- operating a 'radar' of competitive insight within your business sector and beyond, understanding how other organisations respond to major business risks and benchmarking one's performance against these;
- an acute financial awareness of expected growth and revenue, whilst securing a lower average cost of capital;
- an ability to manage foreseeable risk and conditions of true uncertainty by the active monitoring of organisational exposures;
- a capacity to improvise and adapt business models in the face of adversity.

Forward-looking utilities can consider how risk management can support other corporate functions to create value from:

- improved customer, regulatory and investor trust;
- better operational performance;
- more defensible long-term strategies;
- well-evidenced innovation opportunities;
- heightened emergency preparedness;
- better access to finance;
- more effective use of public and investor funds;
- greater employee engagement; and
- stronger, clearer budget decisions and justifications for staffing and resources.

Risks and opportunities can be appraised alongside one another. 'PEST analysis', for example, is a high-level risk and opportunity identification technique that seeks to understand external influences affecting an organisation and, in particular, those factors that affect business continuity and development. It considers political, economic, social and technological factors, and the risks and opportunities to the business that flow from these. In Figure 7.2, this is illustrated for a wood processing company considering relocating its Scandinavian operation to the Czech Republic. The Board-level PEST analysis here (2001) identifies key risks that may threaten business objectives and the opportunities that might be seized. Some practitioners have also added legal and environmental factors to this tool, hence 'PESTLE' analysis in more recent versions.

Political		**Economic**	
Risk	*Opportun ity*	*Risk*	*Opportunity*
• volatile political environment • non-EC country	• Government grants available	• cost of redundancy in Scandinavia • poor infrastructure • supply chain breakdown	• low wages • easy access to finance • skilled workforce
Social		**Technological**	
Risk	*Opportunity*	*Risk*	*Opportunity*
• language barrier • training and education needs • cultural issues	• strong work ethic • good educational system	• poor IT links • low grade technology support	• e-commerce

Figure 7.2 Illustrative PEST analysis for business developments.

Similarly, 'five forces analysis' is a powerful strategic risk identification tool that considers the pressures or 'forces' that influence business, its industry sector and its marketplace. It helps to analyse why a business may be successful or not in meeting its corporate objectives. This approach might equally be used to identify the opportunities associated with these forces. The five forces are:

- the threat of new entrants;
- threat of substitute products or services;
- bargaining power of suppliers or customers;
- competition; and
- intensity of rivalry with the sector.

SWOT (strengths, weaknesses, opportunities, threats) analysis is a familiar management technique that seeks to analyse the strengths, weaknesses, opportunities and threats to a business. The analysis of these four aspects of a particular business decision; e.g. of a decision to embrace a new water or waste treatment technology across a number of sites, can be presented in a similar quadrant to Figure 7.2. Securing business opportunities sometimes requires a shift in mindset from views held historically. Arrival of the stakeholder society has meant that actions conventionally viewed as damaging to business now have tangible benefits in terms of the 'licences to operate' negotiated, not just with regulators, but also by investors and communities (Figure 7.3).

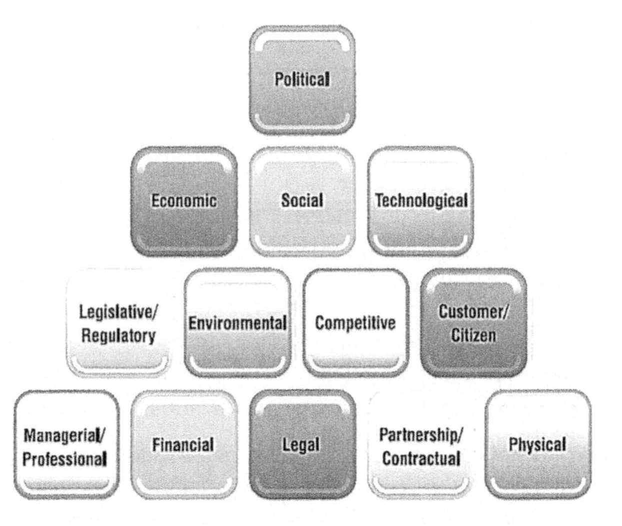

Figure 7.3 Classes of risk and opportunity.

Examples might include:

- securing the benefits of increased investor attention and confidence through corporate disclosures on risk, perhaps historically viewed negatively as 'washing one's dirty linen' in public;
- reaping the rewards of investor interest and public trust through evidence-based triple bottom line (sustainability) reporting, including 'warts and all' assessments, historically viewed as inappropriate disclosure;
- engendering and 'banking' customer trust through wider community engagement in new business developments and projects;
- securing value through cost reduction and the avoidance of misplaced investment using risk-based asset management;
- reducing incident management costs and focused regulatory attention by moving to a preventative risk management strategy; and
- reducing potential exposures from 'contracted-out' activity by driving (and actively monitoring) risk management though the supply chain;

With this mindset (Table 7.1), one starts to view risk management as an opportunity in itself – one that secures the value that might otherwise be lost by poor or misinformed decision-making. One might then argue that the greatest threat to the success of any business is a failure to manage risk – or rather – that the key strategic *opportunity* facing business is the financial, reputational and performance rewards and organisational value that come from sound, effective and transparent business risk management.

Table 7.1 A revised mindset for opportunity management (adapted from Alberts & Dorofee, 2009).

From: conventional risk paradigm	To: opportunity-focused paradigm
Tactical analysis that produces point mitigation solutions	Systemic analysis that produces strategic mitigation solutions
Failure-oriented ("playing not to lose")	Success-oriented ("playing to win")
Narrow trade-off space based on type of risk (e.g. program, security)	Broad trade-off space based on corporate mission and strategic objectives
Applicable to a specific life-cycle phase and a single group, project or team	Applicable across the life cycle and supply chain (multi-enterprise/ system environments)
A stand-alone management practice	Integrated with programme and organisational management practices to maximise value creation
Bureaucratic, 'tick-box' and time-intensive	Practical, straightforward, outcome-focused and easy to apply

7.3 OPPORTUNITY AND PROJECT RISK

We have not dealt with project risk within this text as others devote substantive attention to this specialised activity. However, it is useful to discuss because project risk practitioners have successfully embraced opportunity management (Hilson, 2004, 2006). In part this is because most asset management programmes involve substantive change for complex portfolios of projects and, for a creative programme manager, offer ample opportunity for re-scheduling, cost reduction and programme alteration; all opportunities to save resources and time without compromising the quality of outcome. As a result, many water and wastewater utilities will find staff focused on opportunity management within their asset management teams.

Kähkönen (2001) provides a valuable overview of risk and opportunity management for project managers. Kähkönen observes that seizing opportunities often depends on prompt and well-timed actions executed within an identified 'window of opportunity'. A categorization of opportunities is offered that has much in common with the risk hierarchy and that includes identifying and seizing:

- *business opportunities*, for example project development, customer care during the project life-cycle and attention to high margin activities;
- *systemic opportunities*, typically seizing long term savings such as improved safety, insurance *etc;* and
- *operational opportunities*, for example understanding at the project level which actions are value adding; prioritizing and doing what is important, not compromising safety *etc.*

Kähkönen helpfully illustrates these concepts by reference to the costs of an item as a project progresses, compared to the base cost estimate at the start of the project. The baseline cost estimate (dotted line) comprises commitments of time and cost and the resultant product (situation A) is the base estimate for potential deviations from the expected outcome. In Figure 7.4, situation A demonstrates how the baseline cost estimate for 'item nn' has proved to be too optimistic – the project is under-costed. Here, the three point estimate for minimum, mean and maximum cost shows that even the minimum impact of the main project risk is above the baseline cost estimate, indicating the minimum impact plus a risk premium needs to be added to the target budget for this item. Furthermore, project risk management is required to minimize the final impact of the identified risks in this situation. Here, project risks only result in losses and the margin of safety allowed for this item is insufficient to cover even the minimum impact of the project risks.

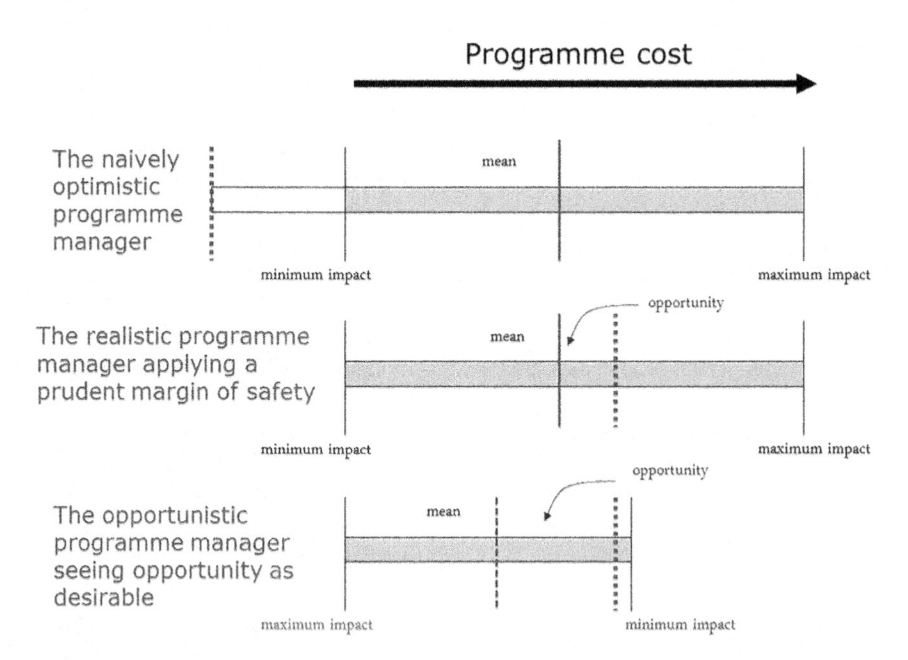

Figure 7.4 Appearances of project risk (A) and opportunity (B and C) relative to the base estimate (after Kähkönen, 2001).

Situation B (Kähkönen, 2001) presents the case where a cost saving (opportunity) has been identified for a new item 'mm'. This is the most widely used approach to managing project opportunities. The baseline cost estimate for item mm is more realistic and includes a margin of safety covering the variance straddled by the minimum and mean impact of the project risks on the item's cost. However, in these situations, the opportunity potential is easily missed, because the situation is so commonly accepted. Concern instead is reserved for the maximum and catastrophic impact. Situation C represents the reverse mindset. We now consider the maximum (positive) impact as the maximum opportunity for savings on our three point estimate, rather than the maximum cost. This shifts the focus and presents the maximum opportunity as desirable. Asset managers in water utilities that manage programmes of investment spend often have this mindset. This capacity to balance and manage risk and opportunity during programme management is a critical skill for optimising the proportion of project budget devoted to project management (Figure 7.5).

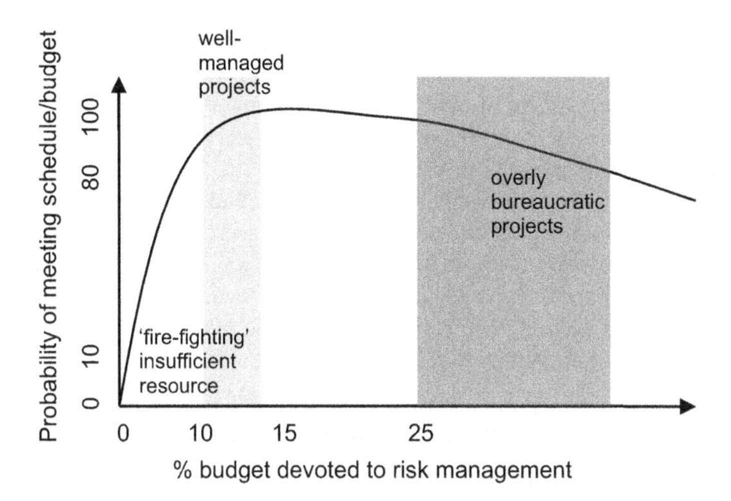

Figure 7.5 Optimizing project management costs (after McConnell, 1998).

With respect to provisions for catastrophic events on the far right hand side of Figure 7.4, a comment on insurance is worthwhile. Clearly, insurers prefer to work with organisations that have well-developed risk management systems and that have integrated risk-based decision making into their management processes. Organisations in high risk sectors without these systems find it difficult to place insurance at reasonable premiums, possibly the most influential source of change in the way that corporate risk management is now being applied arises from the insurance brokerage industry. Brokers are placed between the insured and insurer and are in a powerful position to understand the needs of both parties. Recent changes in the insurance market have led to a hardening of the market and a shift in the risk management strategies of many organisations, some of whom now look to self-insure, (provide contingency for), certain high consequence risks.

7.4 INVESTMENT AND OPPORTUNITY RISK

Behind each strategic investment an organisation considers lies some calculation of the move's worth. Whether considering a joint venture, acquisition, or a major extension to an existing facility, how a utility estimates value is critical to how it allocates its resources, and in turn, is a key driver of overall performance. The most widely adopted valuation framework is the net present value (NPV) model, which estimates value by capitalising, (discounting), future streams of cash flow that the investor expects to receive from an asset. The capitalisation rate is the minimum expected rate of return needed to induce an investor to acquire. Capitalisation is comprised of two components:

- the risk-free rate of return, (accounting for the time value of money); and
- the risk premium, (the additional compensation demanded by investors for assuming risk, illustrated in section 7.3 for project risk).

Risk analysis, in the form of sensitivity analysis and stochastic simulation, is promoted for investment decisions as a means of examining the influence of changes in the key underlying variables on future cash flows, and thus the probability that a project's NPV will fall below zero. Tools that can perform economic evaluation and modelling on a combined entity of investments, (a portfolio), as well as on individual projects are seeing increased application. Luerhman (1997) categorises the valuation of opportunities, (*i.e.* possible future operations) – as distinct from the valuation of current operations. With the former, the decision to invest may be deferred. In 'opportunity valuation', risk matters in two ways:

- the risk of the investment; and
- the risk that circumstances will change before a decision has to be made.

Luerhman describes a common approach in the valuation of opportunities that simply defers valuing them formally until they mature to the point where a decision on whether to seize them or not can no longer be deferred, whereby they can then be valued, in effect, as assets-in-place.

A further category of investment decisions is found where utilities participate in joint ventures, partnerships, or strategic alliances. This takes on particular resonance in the water utility sector, where recent years have seen a proliferation in public/private partnerships. In such cases, where ownership is shared with other parties, managers need to understand both the value of the venture as a whole *and* the value of their company's interest in it.

7.5 MANAGING REPUTATIONAL RISK

Irrespective of their public or private business model, water and wastewater utilities manage what is widely regarded as a public good – water. They provide safe drinking water that has the trust of customers to consumers and treat waste water prior to returning it to the environment. One might consider utilities as being in receipt of three licences to operate:

- From their regulator(s). Regulatory permits and licences that allow the day to day running of the business.
- From funders and investors. Organisational trust in a utility's corporate and business viability allowing periodic capital investment.
- From host communities and customers. Community and consumer trust in the service and in the public interest values of the utility.

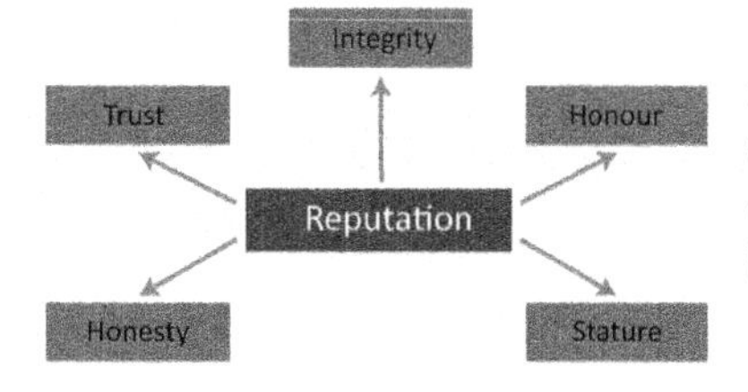

The suspension or loss of any of these 'licences' have significant implications for a utility. Consider the knock-on consequences of the Walkerton outbreak described by Hrudey and Hrudey (2004) beyond the immediate tragedy of 2300 causes of illness and 7 deaths in the Ontario community of 4800 residents, caused by multiple failures in a drinking water system:

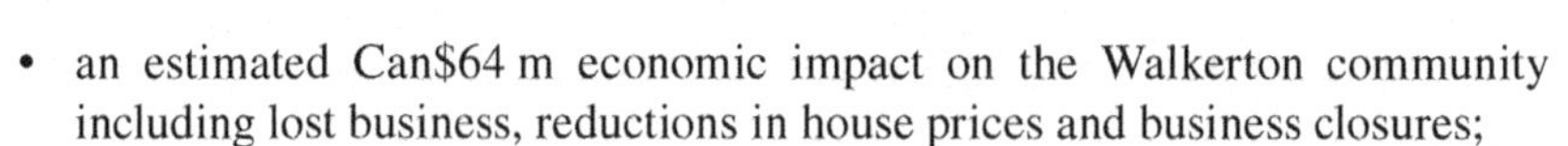

- an estimated Can$64 m economic impact on the Walkerton community including lost business, reductions in house prices and business closures;
- residents experiencing negative treatment in neighbouring communities;
- several hundreds of millions of Canadian dollars expenditure for the Ontario Government; and
- increases in municipal liability insurance premiums across Canada.

Retaining one's licence to operate requires, in part, active management of the beliefs or opinions held by a range of stakeholders about the utility and its operations - management of its reputation, that is. Of the many influences that a strong reputation holds for a utility, access to investment, regulatory relationships and public confidence dominate. As we are often cautioned - trust once lost is very hard to regain. Rayner (2003) describes recent trends in the business climate experienced by organisations in the 21st century:

- the stakeholder imperative;
- globalisation;
- the technological and media revolution; and
- a rise in the importance of intangible assets; especially talent, brand equity and reputation.

With respect to stakeholder trust, good reputational risk management offers a substantive opportunity – the benefits that can be secured when a problem or crisis does occur. Stakeholder trust, 'reputational capital', built up during routine operations, can be of value in the heat of a one-off incident, both in terms of retained consumer confidence and regulatory flexibility during incident management. This is not to suggest it should be relied upon. Utilities need to evaluate and actively manage reputational risk. One's reputation is a construct of the expectations people have about us, amended in the light of their experiences of how we behave. In a business context, reputation is driven by (Rayner, 2003):

- the business track record;
- sound corporate integrity and governance;
- regulatory compliance;
- delivery of a quality service;
- the retention of talented staff;
- demonstrable corporate social responsibility, including environmental performance; and
- communications and crisis management.

For utilities, guarding public and environmental health, communications and crisis management is of particular concern. Together, they represent a key opportunity to test and demonstrate that risk management systems and procedures, risk knowledge and competency training can converge within an organisation under the pressures of an incident. Practitioners of reputational risk management make frequent use of the PEST(LE) analysis technique described above. Sound in a belief that well managed threats can become opportunities and that missed opportunities can become threats, reputational risks and opportunities are often best addressed together. The organisational root causes of adverse reputational risks include:

difficulties in tracking changes in legislation;
inappropriate training or implementation of procedures;
suppliers not adhering to corporate standards;
an absence of an audit function for quality and financial checking; and
poor communication of expectations.

Similarly, among the root causes of reputation enhancement are:

effective marketing;
actively managed media relations;
presence in corporate listings;
elicited customer feedback and response; and
support from non-governmental organisations (NGOs).

7.6 MANAGING EMERGENCIES – AN OPPORTUNITY TO BUILD TRUST

Experience of incidents in the water utility sector shows that without quick authoritative advice and the provision of alternative drinking water supplies, as appropriate, incidents can become high profile events with media and customers driving the agenda during and after the incident. Within four hours of detecting an incident, the utility should be demonstrating it has fully understood key aspects of the situation and is putting actions in place. The rise of social media has escalated the impact and shortened the expectations on reacting to events. Above all utilities must demonstrate a high level of customer care. The objectives of developing a strategy for major incidents are to ensure that:

- the organisation's policy regarding incidents is carried out;
- quick and suitable responses are made to every incident;
- the incident is effectively managed;

- all key actions and duties are carried out; and
- sufficient resources are made available.

EXAMPLE 7.1 MANAGING EMERGENCIES

Emergency management can also be an opportunity to build trust with stakeholders. Good contingency planning is essential for emergency responsiveness. The essential procedures for managing emergencies well include:

- incident categorisation process;
- mobilisation process;
- management structure of the response team;
- identification of the required facilities available;
- definition of each manager's role and duties;
- identification of all communication paths;
- identification of all required communications; and
- identification of emergency equipment within and outside the organisation.

Contingency preparation can be considered a part of water safety planning. Every cause of incidents that could affect the organisation should be identified and the consequences of such a scenario assessed. Examples of possible causes of incidents include:

- deteriorations in water quality
  e.g. pollution of raw water
  failure of a water treatment works
  ingress at a service reservoir
  changes to the operation of the distribution system
- interruptions to supplies
  e.g. source failure
  treatment works failure
  service reservoir failure or isolation
  pumping station failure
  burst main
- incidents at dams or reservoirs
  e.g. increase in water seepage through the structure
  slips or cracking of the dam
  wash out of the dam core
  flooding or erosion
  instability
  over – topping
  landslide into the reservoir
  third party activities
- pollution incidents caused by the water utility
  e.g. discharge of untreated sewage or sludge
  polluting discharge from a water supply installation
  spillage of chemicals or fuels
  escape of toxic gas
- criminal threats or damage
  e.g. bomb or terrorist attack
  kidnapping of employees or their family
  pollution of water supplies
  sabotage

The organisation's policy on responses, together with the standards set by legislation should be agreed by the management of the organisation. These policies and targets may have to be reviewed during the development process if they cannot be reached, are found to be inappropriate or require excessive resources to carry out. Examples of subjects to be covered include:

- response team mobilisation periods
- provision of alternative water supplies
- techniques and timings of communications with the customer
- management of telephone calls from the customer
- communication with the media
- communications with other organisations

Procedures need to be agreed, laid out clearly and circulated to all those nominated to take part. The procedures should include:

- categorisation process
- mobilisation process
- management structure of the response team
- identification of the required facilities available
- definition of each manager's role and duties
- identification of all communication paths
- identification of all required communications
- identification of emergency equipment

The aim of the procedures should be to ensure that all the information is obtained and considered by experts. This will allow decisions to be made on the response based upon the full available knowledge of the situation. Typical titles for the procedural documents are "Major Incident Plan" or the "Incident Response Manual". The response to an incident will depend upon its size, type and impact. To simplify the response procedures it is normal to place incidents into categories, each of which has a stated response. The number of categories and the trigger level of each category will depend upon the structure and objectives of each utility. When the normal management structure cannot cope fully with an emergency without support, then the management required for the emergency needs to be pre-planned. To avoid time-wasting discussion at the start of the emergency, or responding in an inappropriate way, it is suggested that some categorisation benchmarks are agreed. These would be followed at the start of an incident to ensure sufficient resources are available. If subsequently the response had mobilised too many resources, those resources which were not required could be stood down.

7.6.1 Declaration of an incident

There needs to be clear instructions to all employees on the method of declaring an incident. In many industries those instructions need to cover the full 24 hour day. Normally only specified staff can declare an incident, and it is therefore essential that at least one of those nominated is available at any time. Communication and the collection of data to enable the person assessing whether to declare an incident is a vital ingredient the procedures.

7.6.2 Mobilisation

It is important consideration is given to the method and timing of the mobilisation process. During normal working hours, teams can be in place quickly, but at night, weekends or during holidays it will take longer to contact people and for them to make the journey into the centre where they are required. The first three or four hours are the most important in any emergency. If during that period the correct decisions can be made and actions taken, the extent of the emergency and its severity will be considerably reduced. Throughout the mobilisation period of a major emergency (*i.e.* a burst main) it is important that the essential actions are taken. If the emergency concerns a slowly developing situation then at least part of the mobilisation should occur before the situation is too severe.

7.6.3 Key actions of the management team

During a Category 1 incident an Incident Centre is established, manned by the Incident Manager and key members of the response team (Figure 7.6). The role of the Incident Manager is to:

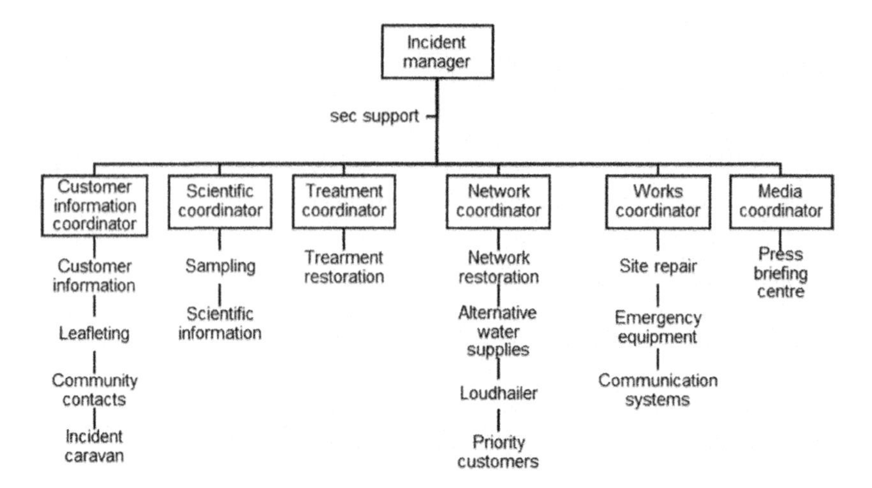

Figure 7.6 Make-up of a typical incident response team with team leader (boxed) responsibilities.

- have overall management of the aspects of the incident
- define the objectives and targets of the response
- implement strategies to meet them
- consider the impact of the incident and the response on the customer, business and the environment
- communicate with other organisations and people (other than individual customers) outside the organisation, media and the press
- manage information given to other organisations, media and customers
- ensure maximum advice and information is given to the customer
- ensure the objectives of these procedures are carried out

The duties of the response team includes:

- identifying and correcting the cause of the incident
- communicating with the customers
- providing temporary alternative supplies
- the restoration of service
- monitoring the effected service

Depending on the type and magnitude of the incident, a number of tasks will have to be performed, and a task leader should be appointed for each one. Examples of these tasks include sampling, alternative water supplies, emergency equipment, transport etc. The responsibilities of the Task Leaders (TLs in Figure 7.6) are:

- obtain the objectives and timescales for the task from the Incident Management Team
- obtain sufficient resources to meet the required objectives
- plan the execution of the duties
- brief all the employees involved in the duties
- monitor progress
- inform the incident management team of progress
- arrange for shift changes where appropriate
- provide welfare facilities for every member of the team

Frequently, incidents can escalate and well defined procedures such as those presented in Figure 7.6 are required. Category 2 incidents will be managed by the Incident Manager and a smaller and appropriate team.

7.6.4 Typical incident management teams for water industry incidents

The management structure and the duties are specifically designed for the emergency situation and may have very little connection with normal structures. For this reason it is better that the roles have different titles to those used on a daily basis.

7.6.5 Nomination

It is essential that employees are nominated for each emergency role included in the procedures document. This will enable the nominees to be trained and take part in rehearsals so that they are prepared for any real situation which may occur. It is often recommended that 3 nominations are made for each post. Two would undertake their duties on a rota basis, as nominees should not be expected to work under emergency conditions for more than 12 hours at a time. The third nominee would be the reserve, which has to be nominated, as there is no standby system because the emergencies will only occur rarely.

7.6.6 Training

The elements of training, which should be given to each nominee, are:

- general appreciation of the procedures, their objectives and each person's role.
- detailed training on the specific nominated role.
- communication skills
- working under stress

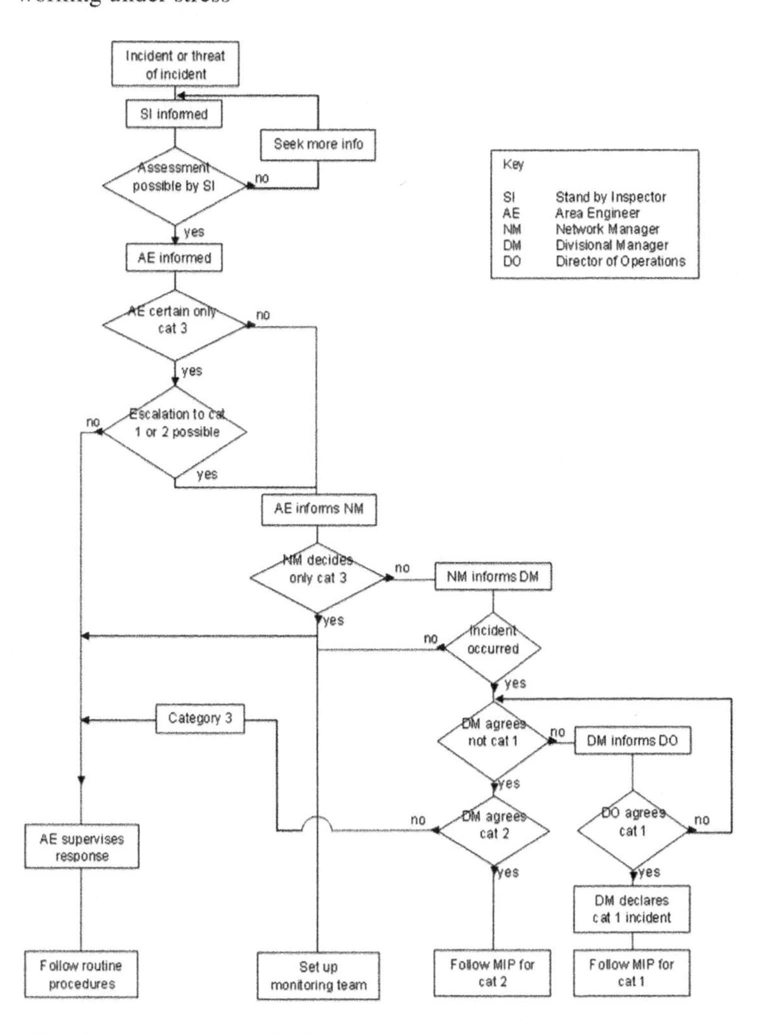

Figure 7.7 Example of an incident escalation process.

7.6.7 Management centres

It is essential to identify the Management Centres that will be used during an emergency. The layout of the room and all the equipment needed during the incident should be identified. Neither the Management Centres nor the equipment has to be retained for the sole use. It can be used for other duties, however it has to be understood by everybody that when an incident is declared, the room and all the facilities are to be dedicated to the Incident Management Team immediately. Facilities such as electric and telephone sockets need to be installed as soon as the Centres are nominated.

7.6.8 Target response times

Target response times, which are measured from the declaration of the incident to the time when the Incident Management Team is operational, should be calculated to give the Incident Manager some idea of the time and hence decisions that have to be made during the mobilisation period. This will depend on the time it takes to contact team members, their travelling time and the time to mobilise the management centres. The response times for working day and out of hours situations will be different. During the response time it is essential that management of the incident takes place and therefore the target response times need to be as accurate as possible.

If necessary tests should be carried out to verify the assumptions. Typical response times are:

- 30 minutes during the normal working day
- 1 hour at all other times

7.6.9 Role of customer call centres

During the normal operation of the business there will be defined methods of handling customer complaints and queries. During a major incident the number of telephone calls will increase substantially. It is very frustrating for the customer and bad for the image of the water organisation if the number of calls being made far exceeds the organisation's ability to respond to them and the customers keep getting the engaged tone. It is therefore important to organise a system which has the ability to receive, handle and process the contents of the high number of calls expected during a major incident. The system design should allow for the transition from a normal to the emergency situation. Frequent briefing should be given to all the team members to ensure they are kept up to date on the changing situation. It is essential that the employees taking the customer calls can understand and relate to their concerns, and can ask questions to update the organisation's information on the incident. This information should be fed to the Incident Management Team through the correct communication channels.

7.6.10 Emergency equipment

A list of available emergency equipment should be drawn up, comprising of equipment that is:

- specifically held for emergencies only
- used operationally
- owned by other organisations

The list should, for each piece of equipment, give details of:

- normal location
- responsible officer
- mobilisation time
- method of transportation
- agreed charges

- training required by driver/operator
- fuels and other items required to operate the equipment
- details of all pipe and cable connections

7.6.11 Contacts directories

A list of all organisations, officers and individuals who may have to be contacted during an incident should be kept up to date, and made available in the Incident Management Room. External communications will typically be required with local authority environmental health, planning and social care personnel (or equivalent), the local health authority including their consultant in communicable disease (or equivalent), the media and the police. Managing and maintaining these relationships outside emergency conditions can prove tremendously productive should an incident occur.

7.6.12 Asset information

In the Incident Management Room should be kept brief details of all the major assets, layout drawings of the treatment works and pumping stations and plans of the trunk main pipelines and the distribution system. This will ensure that basic information is available to the Incident Management Team at the start of the incident. More detailed information can be produced during the emergency but if it starts during the night or at weekends the response time may be slow.

7.6.13 Stand-down procedures

When it is decided to close the management centres, either temporarily or permanently, it is necessary to inform all the people involved in the incident, the press and the customers. Where appropriate instructions should be given on all the actions still to be taken and the targets for progress. There should be a clear and smooth transition from emergency management teams and procedures back to the organisation's normal management structures and procedures.

7.6.14 Post incident review

A full and detailed review of any incident should be carried immediately after every incident. Many organisations use an independent expert to carry out the review and make the results available to their customers to demonstrate open management. The purpose of the review is to:

- fully understand the cause of the incident
- put in place measures to reduce the risk of a similar occurrence
- carry out improvements to the incident response procedures and techniques
- demonstrate where possible to the customers that a high quality response took place and the incident will have no long term effect on health
- consider obtaining additional emergency equipment.

7.6.15 Emergency planning exercises

It is recommended that mock incident exercises involving the Incident Management Teams are held to:

- familiarise employees with their duties
- test the procedures
- identify where additional equipment can be used
- identify where improvements to techniques can be developed

It is recommended that each nominee takes part in an exercise that simulates his role at least once every two years. The exercises should be realistic and based on scenarios which could happen in that organisation. In addition to those exercises, which would involve the management teams, exercises should be carried out to test specific major operations such as the provision of alternative water supplies or the handling of customer complaints. It is important that exercises are reviewed in detail and if changes or training requirements are identified these are carried out promptly.

7.7 SUMMARY AND SELF-ASSESSMENT QUESTIONS

There is an 'upside' to risk that requires active management to the same extent as that applied to the prevention of adverse incidents. Much of the existing expertise in purposeful opportunity management comes from the project management field and, in particular, from the software engineering community. For water and wastewater utilities managing a portfolio of corporate risks, the need is to identify and distinguish between risks that must be managed preventatively and those that might be exploited as opportunities (Table 7.2).

Table 7.2 Deriving strategic benefit from risk management.

Risk Management is not	Risk Management is
Only a means of creating a risk register	A means of helping achieve objectives
Only restricted to the risk process	Part of every business process
Disconnected from organizational objectives	Centred around organizational objectives
Only a management tool for compliance, control and reporting of risks	A tool to drive action
A burden in addition to the 'day job'	A means for reducing the burden of reacting and recovering from failure

Figure 7.8 (Hillson, 2004, 2006; White, 2006a-c) suggests an area of attention (an attention arrow; high risk, high opportunity) for risk managers as they appraise risk and opportunity alongside one another. Because water utilities are often in the public eye, close attention is required to managing the opportunities that arise from sustained performance in the provision of quality water and wastewater services. In our final chapter we consider what constitutes good risk-based decision-making and how an effective organisational risk management culture can be secured.

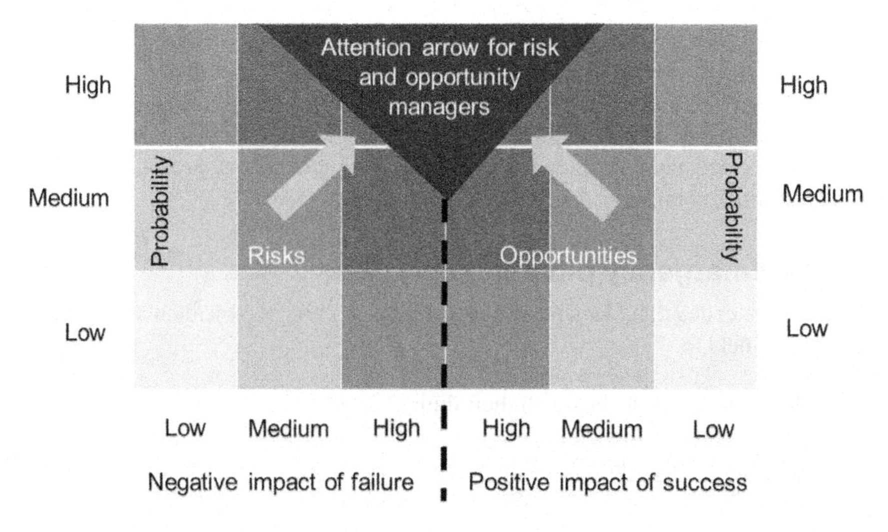

Figure 7.8 Risk and opportunities can be appraised alongside one another (after Hillson, 2004; White, 2006a).

SAQ 7.1 List 4 catchment level risks that can be managed. Now consider what opportunities might be secured from managing these risks.

SAQ 7.2 Undertake a PEST(LE) analysis for the adoption of tertiary (granular activated carbon) treatment of wastewater from a sewage treatment works.

SAQ 7.3 Devise a scheme for categorising water quality incidents. Suggest 3 attributes to defines the incident and then 3 levels of incident – moderate, major and serious.

7.8 FURTHER READING

Alberts C. and Dorofee A. J. (2009). Mission success in complex environments (MSCE) project, A technical overview of risk and opportunity management, Software Engineering Institute, Carnegie Mellon University, available at: http://resources.sei.cmu.edu/library/asset-view.cfm?assetid=21258.

Caldwell J. E. (2012). A framework for Board Oversight of Enterprise Risk. The Canadian Institute of Chartered Accountants, Canada, 69 pp.

Hillson D. (2004). Effective Opportunity Management for Projects: Exploiting Positive Risk. Marcel Dekker, NY.

D. Hillson (ed.) (2006). The Risk Management Universe – A Guided Tour. British Standards Institution, London, ISBN 0 580 43777 9.

Holmes A. (2004). Smart Risk. Capstone Publishing, John Wiley & Sons, Chichester, UK, 270 pp.

Hrudey S. E. and Hrudey E. J. (2004). Safe Drinking Water – Lessons from Recent Outbreaks in Affluent Nations. IWA Publishing, London, UK, 486 pp.

Kähkönen K. (2001). Integration of risk and opportunity thinking in projects. Presented at the 4th European Project Management Conference, PMI Europe, London, UK, 6–7 June, 2001, 7 pp.

Luehrman T. A. (1997). What's it worth? A general manager's guide to valuation. *Harvard Business Review*, **75**(3), 132.

McConnell S. (1998). Software Project Survival Guide. Microsoft Press, US.

Pollard S. J. T and Carroll G. (2001). In: Risk assessment for environmental professionals. S. Pollard and J. Guy (eds), Chartered Institution of Water and Environmental Management, Lavenham Press, ISBN 1 870752 71 6, Suffolk: 21–30.

Rayner J. (2003). Managing Reputational Risk: Curbing Threats, Leveraging Opportunities. John Wiley & Sons, Chichester, UK, 323 pp.

Van Scoy R. L. (1992). Software Development Risk: Opportunity, Not Problem, Technical Report CMU/SEI-92-TR-30, ESC-TR-92-030, Software Engineering Institute, Carnegie Mellon University, 22 pp.

White B. E. (2006a). Enterprise Opportunity and Risk. The Mitre Corporation, Bedford, MA, US, 15 pp.

White B. E. (2006b). Fostering Intra-Organizational Communication of Enterprise Systems Engineering Practices, Draft paper submitted to European Systems Engineering Conference (EuSEC), Edinburgh, U.K., 18–20 September 2006.

White B. E. (2006c). On the Pursuit of Enterprise Opportunities by Systems Engineering Organizations, IEEE International Conference on Systems Engineering, Sponsored by IEEE Systems, Man and Cybernetics Society, 24–26 April 2006, Los Angeles, CA.

Unit 8

Embedding decision making in utilities

UTILITY MANAGER AND BOARD-LEVEL SUPPORT FOR GOOD RISK GOVERNANCE

Utility managers can support their organisations in their risk management journey by adopting a positive approach to the value generated by good risk governance. General managers, with the oversight that comes from involvement in the core strategic plans within a utility, are ideally placed to 'join the dots up' on risk, ensuring that risk management does not become an isolated activity associated with perceived bureaucracy. Best in class utilities that have applied enterprise risk management (ERM) have developed in-house pragmatic approaches that work for their organisational culture.

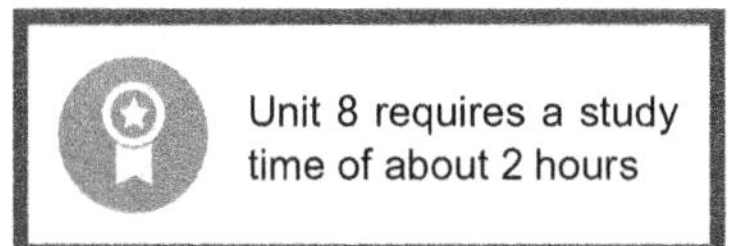

8.1 REFRESHER – WHY MANAGE RISK?

The provision of safe, wholesome drinking water that has the trust of customers is the basic business assumption of a water utility. Preventative risk management is an essential capability for the management of public health risks by utilities because of the latent flaws that develop in all organisations and the fact that no opportunity exists for the 'product recall' of unsafe drinking water – usually the first sign of a problem is disease in the community, by which time, exposure has clearly occurred. This said, the culture of preventative mindfulness essential to managing public health risk is not suited to all risks that a utility faces; for example, without well-informed and measured risk-taking, technological innovation could become stifled in the sector. So utilities have to decide on, communicate and manage a suite of different risk appetites for the full portfolio of risks they manage – this is complex because it requires clarity of communication, clear accountabilities and a culture of responsible risk governance.

So in the 21st century, risk management has become the water utility manager's central function. Beyond the preserves of public health protection and the maintenance of process performance, managers must also assess and manage legal, financial, environmental, resource, technological and reputational risk, among others, in an increasingly complex and integrated business climate. The implications of their decisions and accountabilities stretch well beyond the company structure to publics, regulators, consumers and shareholders alike. Matched to this has come an increasing degree of scrutiny of the underlying basis, and processes for, water utility decision-making. Managing risk within the business environment is complex and difficult. It requires an organisational capacity to:

- anticipate and assess risk at the strategic, programme and project/operational scales (from issues as diverse as skills retention, to water safety planning and maintenance scheduling);
- meaningfully compare and prioritize risks of widely varying characteristics (Figure 8.1);
- distinguish between simple risk tools and more sophisticated methods;
- manage risk reduction without unduly compromising business competitiveness;
- set in place practical mechanisms for risk identification and management;
- prioritize issues for immediate action and develop contingency procedures; and above all
- develop a risk aware culture of proactive risk management rather than risk 'avoidance'.

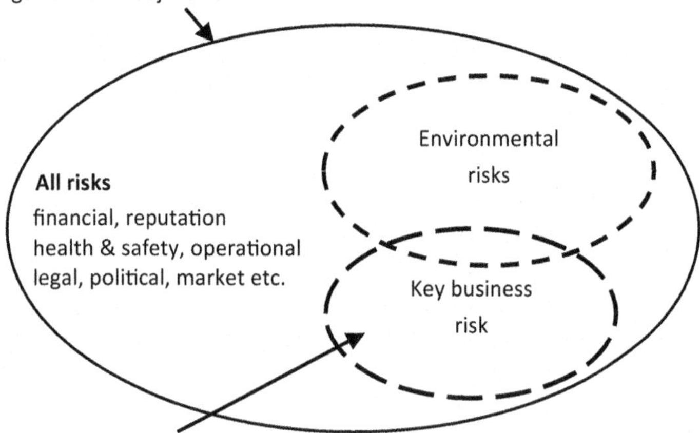

Figure 8.1 Relationships between all risks, key business risks and environmental risks for an organisation.

The essential features of good risk management (O'Connor, 2002) are:

- being preventive rather than reactive;
- distinguishing greater risks from lesser ones and dealing first with the former;
- taking time to learn from experience; and
- investing resources in risk management that are proportional to the dangers posed.

Risk management must aid business objectives and not be a brake on progress, innovation or the seeking of business opportunities and new ventures. A key goal of good corporate risk management then is to make organisations more efficient and customer responsive:

- *efficient* in that resources (finance, staff time and levels of expertise), are used wisely on key areas of business activity such that risks are actively managed rather than merely creating an illusion that a risk has been 'avoided'; and
- *customer responsive* in that utility decisions can be made and communicated clearly, even in the face of complexity and uncertainty by Boards, managers and customer service functions.

8.2 SECTOR PROGRESS IN RISK MANAGEMENT

Achieving wholesome, affordable and safe drinking water that has the trust of customers means, as a minimum, that water is safe in microbiological and chemical terms; acceptable to consumers in terms of taste, odour and appearance; and that the supply is reliable in terms of quality and quantity (Figure 8.2). Delivering these objectives within a multi-stakeholder, institutional and business context in which expectations are rising requires:

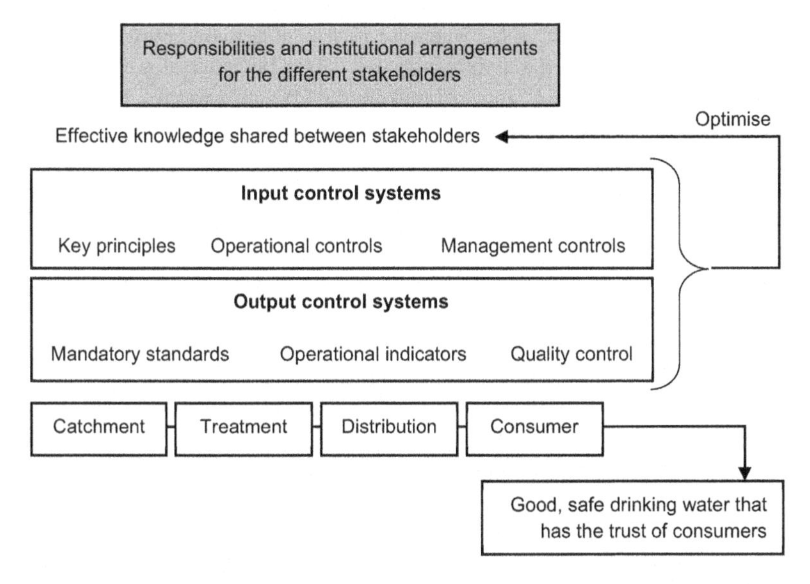

Figure 8.2 Proposed framework for the assurance of water quality (after AWWA *et al.* 2001).

(i) the management of risk from source to tap (water safety plans);
(ii) an integrated approach with co-operation from water suppliers, Governments and their agencies, land users, contractors and consumers (shared risk management responsibilities); and
(iii) transparency in the maintenance of quality assurance and decision-making (openness, beyond compliance, communication with others).

The debate on risk has crystallised the relationship between risk management and organisational performance. Critical aspects include:

- the importance of openness, transparency, engagement, proportionality, precaution, evidence and responsibility to good decision-making;
- the value of taking a long-term perspective in assessing the potential consequences of management actions; and
- recognition of a widely-held experience that 'hard' quantitative risk analysis tools used in isolation of transparent decision-making does little to gain public confidence and can result in the long-term erosion of trust.

Particularly timely is the risk management philosophy underpinning revisions to the World Health Organization's (WHO) Guidelines for Drinking Water Quality (3rd and 4th revisions). These place an emphasis on the development and implementation of water safety plans (WSPs) for water quality management and, within these, the application of risk frameworks such as the "hazard analysis and critical control points" (HACCP) approach – adapted to water quality management from the food processing sector and promoted as a basis for prioritizing risk management measures within the water supply chain from catchment to tap.

Most utilities now have structures in place for business risk management. Typically, there are split accountabilities for risk management within a utility management structure such that the Chief Financial Officer/Financial Director and Board have overall responsibility, supported by an internal audit function for the management of strategic risks; middle management for programme level risks, (e.g. asset management, resource planning), and individual project staff for operational risks (e.g. plant performance). Many of the larger utilities have in place risk management committees and group risk managers that monitor and report on priority risk areas within the utility, principally in response to requirements on internal (financial) control. Beyond these company boundaries, there are important risk interfaces with other institutions including regulators and the capital markets.

One such interface concerns the financing of infrastructure spend and subsequent asset management. Over and above the levy of rates at the local level to support investment on the maintenance of water and wastewater treatment systems, infrastructure spend within public sector utilities is supported through the tax base and granted via a series of Government aided revenue streams or funds. In the US, for example, the Clean Water Act State Revolving Fund (CWASRF), provides financial support to individual states to pay for sewage treatment works ($1.35 billion in FY 2002); and the Safe Drinking Water Act State Revolving Fund supports capital investments in water treatment facilities.

8.3 TOOLS AND TECHNIQUES

There is now a considerable literature on the application of risk techniques for evaluating and characterising the various categories of risk and an increasing expectation of their application within the regulatory context.

Recognise, however, that risk analysis plays a role alongside other decision tools for risk management. Not all risks require detailed analysis to be managed. In many industries there are accepted engineering standards of performance and codes of practice which, if adhered to, provide high degrees of control with the confidence that comes from good engineering design and accepted control procedures. These are applied where uncertainties and system vulnerabilities are well understood. However, complex, uncertain and novel systems, and situations where there is a deviation from routine operation require prior risk analysis, to better understand what drives the risk from or to the plant, process or operation, thereby allowing management measures for the reduction of unacceptable risks to be implemented.

Risk assessment has long been the basis for the derivation of water quality standards for drinking water and in this context is widely accepted within the sector. Though critical to the derivation of water quality guidelines and the protection of public health, these substance-specific health risk assessments are, in practice, somewhat distanced from the immediate operational context of individual utilities, sitting behind the water quality guidelines that they generate. Applications of risk analysis within the operational context can be illustrated by the assessment of risks from hazards such as

- *Cryptosporidium parvum* in surface and reclaimed waters;
- pollutant transport and exposures within sewer systems;
- pathogens in untreated wastewaters;
- excess nutrient loadings and of exotic chemicals (e.g. pharmaceuticals); and
- the failure of automated alarms on water treatment plant.

In this text we have described the assessment and management of risk at the strategic, tactical (programme level) and operational levels of a utilities business (Table 8.1).

Table 8.1 Risk analysis strategies for operational, programme level and strategic risks.

Level	Example	Typical Risk Analysis Approach and Tools
Operational	Plant reliability and performance; individual compliance assessment	Failure mode and effect analysis; fault tree analysis; probabilistic analysis based on historic discharge data
Programme level	Watershed/catchment analysis, e.g. pathogens; asset management – failure of service levels; CSO and intermittent discharges; regulatory assessments of discharges to rivers	Environmental distribution models and exposure assessments; semi-quantitative risk ranking tools; systems level tools for well characterized distribution networks;
Strategic	Investment, acquisitions and investment strategies	Financial cost-benefit models with value-at-risk add-ons;

A key operational element of risk analysis is the judgement of evidence, particularly as it relates to water quality and public health risk. Hrudey and Leiss (2003) have provided insights to the limitations of routine monitoring for providing warnings about what are in essence rare hazards; and about the futility of pursuing caution beyond what our knowledge and monitoring systems can support. These insights provide a practical perspective on how monitoring systems need to be designed and evaluated to provide the most useful understanding of risks so that appropriate judgements can be made in balancing between type 1 (false positive) and type 2 (false negative) errors. Maintaining an appropriate balance in this regard is the essence of sound risk management decision-making.

Of additional relevance to the operational context are probabilistic risk assessments that evaluate the risk of non-compliance with permits, either for failing or newly-commissioned plant, and the risk-based tools used by regulators to assess the potential threats to water supplies from ongoing, or accidental releases within catchments.

Moving on to the programme level, we are concerned primarily with risk techniques used to evaluate the risks posed by an identical hazard experienced across the business functions of a utility (e.g. a new IT 'roll-out') or which is experienced at a variety of locations (e.g. mains bursts, progressive failure of filter media – in asset management, for example) or with the wide variety of risks existing within a watershed. The availability of geographic information systems has facilitated the ease by which such assessments can be performed. Programme level risk assessments are concerned with the implementation of strategies across multiple sites (e.g. the 'roll out' of strategies and programmes within utilities) and geographic regions (catchment/watershed planning). In the US, for example the US Natural Resources Defence Council (2003) reported on the risk to drinking water quality from aging pipes and process plant across the US with individual city 'rankings' being informed by water quality data, USEPA compliance records and water utility annual reports. Similar risk ranking techniques have been used internationally to inform threat assessments of water systems in light of recent terrorist activities.

In Europe, the DPSIR approach (unit 4) to identifying key hazards within a watershed, by reference to the driving forces (e.g population growth), pressures (sewer discharge), state (increased nutrient load), impacts (anthropogenic eutrophication) and policy response (discharge control) is being adopted under the European water framework directive (IMPRESS, 2002). Here, risk assessments will inform a programme of activities targeted at raising the ecological status of the watershed. Given the plethora of potential catchment management issues in any programme, there is a need to prioritize risk management within the watershed by concentrating on those measures that reduce the significant likelihood of severe impacts being realized.

Programme level risk analysis invariably involves trading costs and benefits. Typical are the risk-based resourcing approaches briefly described above. When designed well, piloted and implemented with feedback, these systems can provide a sound basis for distinguishing greater risks from lesser ones, and for investing resources in risk management that are proportional to the risks posed. However, these systems, whether to drive maintenance schedules, monitoring regimes or workforce planning, may also incur risk unless the consequences of resource trade-offs are fully understood.

At the *strategic* level, financial and project risks associated with infrastructure investment, merger and acquisition activity, outsourcing arrangements and the long term viability of investment decisions are the key concerns. For example, the US Clean Water Act and Safe Drinking Water Acts are requiring investments of $23 billion over the next 20 years to upgrade water infrastructure. Risk analysis is critical to optimising and targeting the spend of such vast sums.

8.4 IMPLEMENTING RISK MANAGEMENT

Embedding risk management deep within a utility, however, is more than having a risk framework, a suite of decision tools and company risk champions in place –

though these are important factors. It requires changing hearts and minds so that responsibilities are shared, corporate knowledge is well managed and essential communication channels, both within and between organisations, are kept open and functional in the event of an incident.

Making credible and defensible decisions in organisations requires an institutional capacity to be preventative rather than reactive when managing risk and an aptitude to capture and learn from experience. Implementing risk analysis strategies and decision-making frameworks requires clear, straightforward procedures that can be understood, agreed and operated by all levels in an organisation – indeed, 'keeping it as simple as it needs to be' has been a mantra of expert risk analysts. Risk mature organizations have the following key processes in place:

(1) **Core risk management processes**, including: (i)setting and allocating risk and reliability requirements; (ii) performing risk analyses including reliability studies to inform decisions; (iii) design and operation of plant to meet specified risk and reliability requirements; and (iv) risk assurance to the customers, stakeholders and regulators.
(2) **Organisational implementation processes** including: (i) the verification of management processes and validation of risk models and data; (ii) project risk management ,(ensures risks managed to cost and delivery schedules); (iii) emergency response management; (iv) reliability, qualification and safety testing (provides assurance of performance); (v) measurement and analysis of data (assurance of what is achieved in service); (vi) procedures for the management of change (identifying key differences); (vii) supply chain management (sometimes failures and incidents have their causes in products supplied down the supply chain).
(3) **Institutional support**. Finally, there is an expectation that the implementation of a risk management framework will be supported by organisational learning, education and training programmes and commissioned research and development.

Implementing risk management is challenging because utilities seek to engender an intelligent approach to managing risk and yet keep messages straightforward and uncluttered with risk jargon – messages that proactively protect public health and, in their implementation, avoid a 'tick-box' mentality under which there could be a danger of becoming disengaged. Driving this agenda forward practically and, at the same time being willing to take calculated risks to achieve overall business gain, provides a common motivating interest for many utilities.

Ultimately, good risk management in the water and wastewater utility sector involves the anticipation and management of change. We know that many water quality incidents follow periods of change in natural or human processes. Following the types of failure reviewed by Hrudey and Hrudey (2004), it is accepted that risk analyses need to extend their reach beyond engineered systems and view management (system) and human (people) factors as equally central to effective risk management. Risk managers also know that risk assessments in isolation do not guarantee risk reduction. Left with their recommendations not implemented, they are a hollow gesture. The frameworks that exist to guide risk managers then are valuable up to a point, but remain somewhat at a distance from the process of managing risk on the ground in many organisations.

8.5 SECURING A RISK MANAGEMENT CULTURE

The recent emphasis on risk governance in water utilities shows that sound risk governance generates value, confidence in others, 'no regrets' decisions and a 'trusted partner' status. Those utilities that have put in place organisational structures and developed capabilities in risk management recognise that managing risk, resilience, adaptation and opportunity is an organisational journey requiring a responsive culture with agreed accountabilities at all levels. Developing a risk management culture that is sustaining and continues to learn and improve in the face of the inevitable peaks and troughs of organisation performance requires:

- leadership;
- procedures;
- an appetite for conservative decision-making where safety is put first even under pressure;
- a culture of sharing reported near misses;
- good communication at the appropriate level;
- an open, learning organisational culture open to benchmarking against the 'best-in-class';
- systematic competency checking;
- effective management of organisational change; and
- the ability to prioritise.

In practice, our utility systems are not simple reliability block diagrams of three or four unit processes in series. They are complex networks of interconnected assets and sub-assemblies in a constant state of stress that comes from their operation. As a result, flaws lie deep within them, so we should expect there will be failures and so be mindful and root flaws out before they contribute to incidents. The management literature is rich in the forensic analysis of accidents. Disasters have deeply-rooted causes that are a combination of technical failures, an incapacity to manage change and of the underlying values within, or market forces acting on, an organisation. Accidents typically occur (Taylor, 2005):

- when there is a loss of institutional foresight and corporate memory;
- in the face of strong market pressures for efficiency gains;
- when there are considerable elements of outsourcing;
- where organisations fail to maintain their status as an 'intelligent customer';
- with loss of internal technical expertise and particularly during, or following periods of business process re-engineering.

Cost pressures, priority-based working and changes that are rushed are all circumstances that can generate accidents. For organisations to become resilient, they must anticipate and circumvent threats to corporate objectives and manage pressures and conflicts between positive business performance and the risks that threaten it.

Modern management culture sets a strong impetus on doing more for less and on maintaining business continuity. Middle managers may find it difficult to challenge this philosophy in elevating risk issues to the executive or Board. When they do, risks that are not easily quantified in monetary terms may receive less attention at risk register review meetings, potentially remaining unaddressed and in the worst case, lie dormant within the organisation as latent precursors for future accidents. Leadership and management are key to establishing the right culture in terms of the expectations and example that are set, or not. This all sounds like common sense, but many practical challenges exist:

- How do organisations develop a risk management culture without having first to suffer a major accident?
- How do we force ourselves to ensure risk issues are treated seriously?
- How can we usefully process the volumes of risk information gathered by risk managers so as to make sense of it for accident/incident prevention?
- When should executive managers listen to the challenge from below?
- Above what threshold should they act?
- Are risk managers an additional source of risk because, in taking institutional responsibility for coordinating risk assessment and management, they absolve others of their individual responsibilities for risk management?

These are critical organisational questions germane to the organisational practice of risk management and requiring research. They remind us that managing risk requires wisdom and reflection, and preventive approaches that are creative and forward-looking. Whilst zero risk is not achievable; vigilance and progressive improvement is.

CHECK POINT FOR UTILITY EXECUTIVES. HAVE YOU:

- Appointed a focal point for coordinating risk management efforts in your utility and supported them with requisite resources?
- Given them access to the Executive to ensure they understand the strategic agenda, with opportunities for regular risk reporting and informal discussion?
- Considered a systems view of your utility that extends beyond the physical assets, identifying other critical points of control to ensure risks are well managed?
- Reflected on what, in your utility, will secure a pervasive and preventive approach to risk management without compromising your ability to be agile and to secure business opportunities?
- Reinforced what you need of others, with respect to risk management, at recent management briefings?
- Discussed risk appetite at a Board level, and then communicated it to senior managers for onward cascade to employees? Are employees aware of the risks you are willing to have them take in their roles and those they should escalate to more senior levels? Have they a means of doing this?
- Are you prepared to splice a robust risk governance philosophy into the DNA of your organization?

8.6 CONCLUSIONS – HIGH RELIABILITY AND 'MINDFUL' ORGANISATIONS

What might we learn from other sectors with public safety responsibilities? The management literature suggest some organisations operate in highly complex and hazardous sectors but almost never experience a high impact failure. This diverse group of organisations are referred to as high reliability organisations (HROs). HROs include members of the nuclear, chemical and aerospace industries. These organisations operate 'high-hazard, low risk' technologies at extremely high levels of reliability. Creating high reliability requires vigilant managers reacting to unexpected triggering events by putting in place organisational structures that help, trap or mitigate against cascading error.

The phrase 'mindful organisations' has been applied to organisations successfully operating with high reliability. Best in class organisations are mindful about risks to their operations. Weick and Sutcliffe (2001) characterise mindfulness in organisations that:

(i) are preoccupied with failure and the root causes of it (Swiss cheese model);
(ii) are reluctant to (over)simplify (recognise complexity);
(iii) are sensitive to operations (understand the role of change to risk);
(iv) committed to resilience (they seek systemic changes that build robustness); and
(v) are deferential to expertise (deep expertise trumps organisational status).

Securing 'mindfulness' for the water utility sector is critical to the implementation of water safety plans. It is vital that water companies proactively seek out new hazards, not just offer analysis to support current practices. A key factor in the development of an organisation's capacity for mindfulness is the management ethos and regulatory and legal framework in which it operates.

It is recognised that a 'just' culture within organisations is required for an open and effective communication and management of risks. Organisational cultures are deemed to comprise elements (Johnson, 2002; Content, 2005; Figure 8.3) of the:

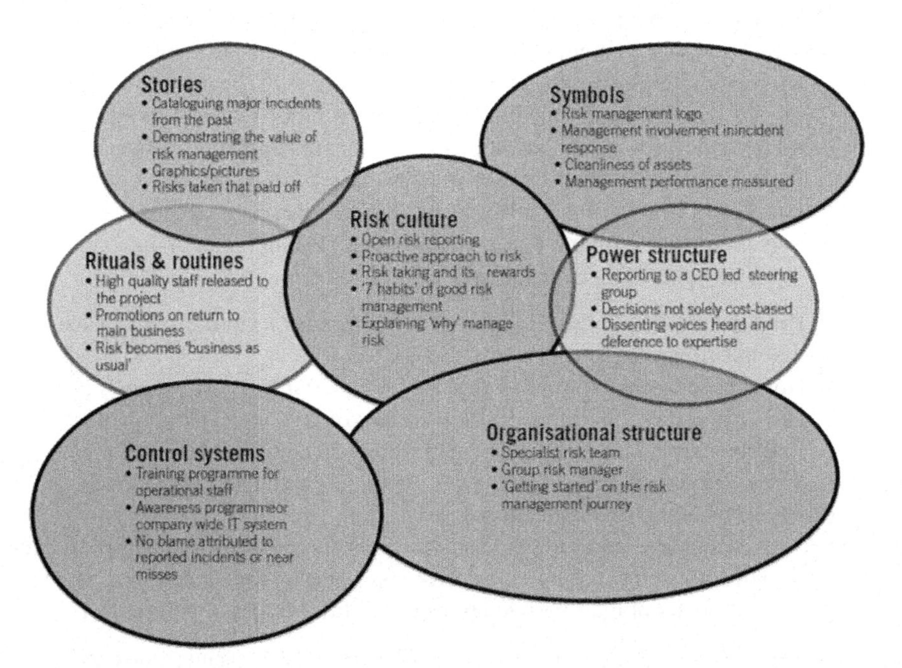

Figure 8.3 A cultural web for risk management with selected features (Content, 2005; adapted from Johnson, 1992).

- **rituals** of organisational life, such as training programmes, promotions of what is most important in the organisation, reinforcement of "the way we do things around here" and signals of what is especially valued;
- **stories** told by members of the organisation to each other, to outsiders, to new recruits and so on that embed the present in its organisational history and flag up important events and personalities, as well as mavericks who "deviate from the norm";
- **symbolic aspects** of the organisation such as logos, offices and titles or the type of language and terminology commonly used: these symbols become a short-hand representation of the nature of the organisation;
- **control systems**, measurements, and reward systems that emphasize what is important in the organisation, and act to focus attention on particular systems;
- **power structures**: the most powerful managerial groupings in the organisation are likely to be ones most associated with core assumptions and beliefs about what is important; and
- formal **organisational structure** and the more informal ways in which the organisation works are likely to reflect power structures and again, delineate important relationships and emphasize what is important in the organisation.

Much has also been made of the role of risk knowledge and information. IOSH (1994) are of the view that the first requirement for the generation of a risk culture is to create a reporting culture, though a prerequisite for this is the capacity to manage knowledge. For a water utility, this might include the establishment of:

- regular risk management meetings to review controls and actions on significant exposures;
- a risk and issues database with established scoring protocols;
- procedures for escalating significant new risks to directors;
- appointed roles of risk champion and risk coordinator in each directorate – the former are senior managers with the experience to test, challenge and improve the efficiency and effectiveness of the risk management process; risk coordinators implement the risk management process.

8.7 SUMMARY AND SELF-ASSESSMENT QUESTIONS

Unless risk assessments are enacted upon, they are worthless. Managing risk requires corporate structures, accountabilities and processes. Beyond this, organisations must also develop proactive, preventative risk management cultures. This is not straightforward, nor can it be achieved especially quickly. However, the prize is substantial. Organisations that are in tune with their business risks and opportunities can optimise processes without putting systems at unnecessary risk; their employees identify issues before they escalate and their stakeholders and investors are prepared to demonstrate their confidence by imbuing trust and investment.

SAQ 8.1 What do you understand by the term 'preventative risk management'? Provide an example for each of the strategic, programme (tactical) and operational contexts of water utility management.

SAQ 8.2 Why is knowledge management important to effective risk management?

SAQ 8.3 What kinds of company 'stories' might be useful in helping generate a risk management culture? Can you devise one? Ensure is includes 3 lessons learnt.

8.7 FURTHER READING

Aon Corporation (2007). Enterprise Risk Management: Practical Implementation. Aon Global Risk Consulting, Aon Corporation, Chicago, IL.

ACCA (2011). Rules for risk management: Culture, behaviour and the role of accountants. Accountants for business. The Association of Chartered Certified Accountants, December 2011.

Australia/New Zealand Standard (1999). Risk Management Standard . AS/NZ: 4360:1999. ISBN 0 7337 2647 X.

Awwa, EUREAU and WSAA (2001). Bonn Workshop 2001 – Key Principles in Establishing A Framework for Assuring the Quality of Drinking Water: A Summary. AWWA, 6 pp.

Awwa Research Foundation (2007). Risk Analysis for More Credible and Defensible Utility Decisions. Awwa Research Foundation Project 2939, Awwa Research Foundation, Denver, CO.

Bradshaw R., Gormely Á., Charrois J. W., Hrudey S. E., Cromar N. J., Jalba D. and Pollard S. J. T. (2011). Managing incidents in the water utility sector – towards high reliability? *Wat. Sci. Technol: Water Supply*, **11**(5), 631–641.

Canadian Institute of Chartered Accountants (2003). 20 questions directors should ask about risk (ed. H. Lindsay), Canadian Institute of Chartered Accountants, Toronto, ON, 12 pp.

Carter M. (2012). Safety Critical. Presented at the Project workshop: Risk governance – a water utility manager's implementation guide, 7–8 March, 2012, London, Cranfield University, Water Research Foundation and UKWIR, UK.

Cass Business School (2011). Roads to Ruin. A Study of Major Risk Events, Their Origins, Impacts and Implications. Prepared for Airmic, London, 184 pp.

Content N. (2005). Securing Corporate Buy-In to Business Risk Management, UK presented at the AwwaRF International Workshop "Risk Analysis for better and more credible decision-making", Banff Centre, 6–8 April, 2005, Banff, Alberta, Canada copyright 2005 Cranfield University and American Water Works Association Research Foundation.

CSA (1997). Risk Management: Guideline for Decision Makers -Q850-97 (R2009) Canadian Standards Association.

Deal T. E. and Kennedy A. A. (1982). Corporate Culture: The Rites and Rituals of Corporate Life. Addison-Wesley, Reading, MA.

Dunn A. J., Frodsham D. A. and Kilroy R. V. (1998). Predicting the risk to permit compliance of new sewage treatment works. *Water Sci. Technol*, **38**(3), 7–14.

Economist Intelligence Unit (2011). The Long View: Getting New Perspective on Strategic Risk. Economist Intelligence Unit Limited, London, 35 pp. from London@eiu.com.

ECPB (2011). Free Risk Management Guide, Emergency Preparedness Capacity Builders (EPCB), Adelaide, South Australia http://www.users.on.net/~salters/Risk%20 Management%20 Guide_1.1.pdf.

Fewtrall L. and Bartram J. (2001). Water Quality: Guidelines, Standards and Health – Assessment of Risk and Risk Management for Water-Related Infectious Disease. WHO, IWA Publishing, London.

Fraser J. R. S. and Simpkins B. J. (2005). Ten common misconceptions about enterprise risk management. *Journal of Corporate Finance*, **19**(4), 75–81.

Frigo M. L. and Anderson R. J. (2011). Embracing Enterprise Risk Management – Practical Approached for Getting Started. COSO, January 2011. Durham, NC.

Hellier K. (2000). Hazard analysis and critical control points for water supplies, Proc. 63rd Annual Water Industry Engineers and Operator's Conf., Warrnambool, 6–7 September, 2000: 101–109.

Hrudey S. E. and Leiss W. (2003). Risk management and precaution: Insights on the cautious use of evidence. *Environ. Health Persp.*, **111**(13), 1577–1581.

Hrudey S. E., Hrudey E. and Pollard S. J. T. (2006). Risk management for assuring safe drinking water. *Environ. Intl.*, **32**, 948–957.

Institution of Occupational Safety and Health (1994). Policy Statement on Safety Culture. IOSH, Leicester, UK.

Jalba D. I., Cromar N. J., Pollard S. J. T., Charrois J. W., Bradshaw R. A. and Hrudey S. E. (2010). Safe drinking water: Critical components of effective interagency relationships. *Environ. Intl.*, **36**, 51–59.

Johnson G. (1992). Managing strategic change – strategy, culture and action. *Long Range Planning*, 1992, **25**(1), 28–36.

National Health and Medical Research Council (2001). Framework for Management of Drinking Water Quality, NHMRC, New Zealand, 71 pp.

NRDC (2003). What's on Tap? Grading drinking water in US cities, Natural Resources Defense Council, Washington DC, 226 pp. at: http://www2.nrdc.org/water/drinking/uscities/pdf.

O'Connor D. R. (2002). Report of the Walkerton inquiry. Part 1. The events of May 2000 and related issues, The Attorney General of Ontario, Toronto, The Walkerton Inquiry.

Tranfield D., Denyer D. and Marcos J. (2002). Management and development of high reliability organisations. *Management Focus*, **19**, School of Management, Cranfield University, Cranfield, UK.
UKOOA (1999). Industry Guidelines on a Framework for Risk Related Decision Support. UK Offshore Oil Operators Association, London, 32 pp.
Weick K. E. and Sutcliffe K. M. (2001). Managing the Unexpected – Assuring High Performance in An Age of Complexity. University of Michigan Business School, Josey-Bass Publ., San Francisco, CA.
WHO (2004). Water safety plans, Chapter 4. WHO Guidelines for Drinking Water Quality, 3rd edn. Geneva, World Health Organization, 2004, 54–88.

Unit 9

Summary

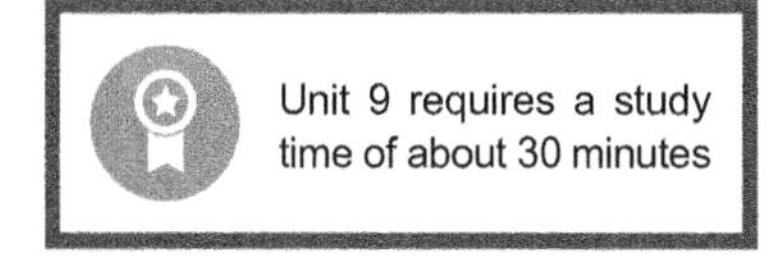

9.1 MANAGING CHANGE IS RISK AND OPPORTUNITY MANAGEMENT

Managing risk well is a key competency for water utilities, and many utilities have established risk manager roles (Figure 9.1) to coordinate their efforts. An essential requirement for utilities is to develop a preventative and anticipatory approach to risk and opportunity that ensures they are resilient to threats, whilst equally alive to opportunities. In practice, this means developing an organisational capability to connect operational activities to utility-wide risk management programmes; to understand the impact of risk on a utility's corporate priorities; and then forecast future risks into the mid- and long term so stakeholders can be confident in the master plans that utilities develop with others to manage risk over the planning cycle.

Figure 9.1 A typical professional journey for a risk manager new in role within a utility (after Fraser, 2005). The figure charts the activities forward in time from appointment, assigning organisational value to the various activities in each 6 month time slice from bottom to top. Note the strategic value after 2–4 years in post of external activity aligned to the needs of external stakeholders.

Credit: istockphoto © AvailableLight

For water and wastewater utilities, risk management has been their main principal business for 150 years (Figure 9.2). The provision of safe drinking water, through the operation of available, reliable and maintainable unit processes for the removal of pathogens and undesirable chemicals is one of the key engineering and public health successes of the 19th and 20th centuries. Why then, this refocusing of effort on risk management, given the track record? One argument relates to the pace of change. Often with change comes uncertainty and the opportunity to perturb previously stable systems. Examples of substantive change within the sector include:

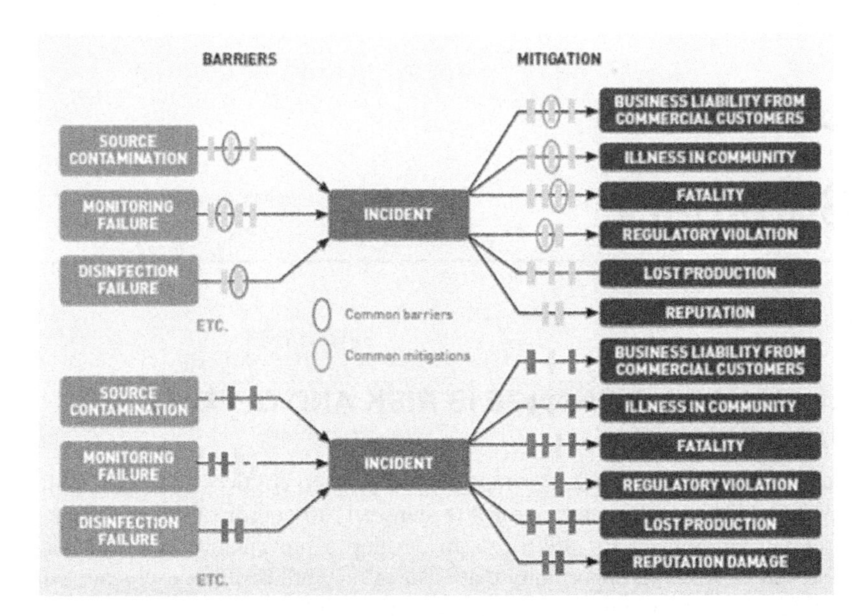

Figure 9.2 The core capability for any water utility must be the maintenance of effective barriers against contaminants, especially pathogens, that may enter the public supply. The bow time approach illustrated here (after Carter, 2012), summarises barrier integrity and the status of risk management actions designed to mitigate against the undesirable consequences on the right side of this figure. This mindset, of vigilance and mindfulness against threats to barrier integrity is an essential competency of modern water and wastewater utilities.

- a progressive deterioration of Victorian and 20th Century infrastructures;
- altered governance arrangements for the management and financing of utilities;
- increased regulatory requirements, including pressures to remove and treat hitherto unforeseen hazardous agents, both pathogenic and chemical;
- changes in employment patterns, with increased staff turnover internally within organisations and between organisations, coupled with the perceived loss of long-standing corporate knowledge;
- increased stakeholders expectations of corporate bodies, including high public demands for safety and reliability in concert with public anxiety about environmental chemicals;
- a perceived recent decrease in global security.

9.2 ORGANISATIONS THAT ARE MATURE IN RISK MANAGEMENT

Understanding, or more properly, preparing for the implications of these and other changes requires an organisational capacity to evaluate the potential consequences associated with change – a capacity to foresee and act in advance of hazards becoming manifest. This the key requirement for utility managers, process engineers, water quality scientists, regulators and contractors introduced in this text – developing an active risk management mindset. What are its features?

You know when your organisation has developed risk management maturity when:

- the results of risk assessments are being implemented and risk assessments are being periodically revisited;
- people are visibly rewarded for having foreseen and prevented accidents;
- your key business risks are communicated to staff and staff understand their daily contribution to managing these;
- the organisation's corporate values are visible – this is critical because it sets the very tone of risk management. Utilities charged with public health protection must demonstrably pursue this goal;
- you drive risk management through your supply chain and outsourced business functions;
- you are progressing a competency programme for staff in risk assessment and management; and when
- you are prepared to disclose the key challenges to your organisation and report on the progress you have made towards managing these risks.

9.3 MINDFULNESS FOR THE WATER AND WASTEWATER UTILITY SECTOR

The provision of safe drinking water deserves to be treated as a 'high reliability' service within society and subject to the sectoral and organisational rigours and controls inherent to operations in the nuclear, offshore and aerospace industries. These sectors have learned important lessons and developed significant literatures on the implementation of safety cultures, much of which is transferable directly to the water sector as it progresses with the implementation of risk management. Weick and Sutcliffe (2001) characterise 'mindfulness' in organisations that (i) are preoccupied with failure and the root causes of it; (ii) are reluctant to (over) simplify; (iii) are sensitive to operations; (iv) committed to resilience; and (v) are deferential to expertise. In conclusion, it is proposed that for water utilities seeking to develop mindfulness (Hrudey *et al.*, 2006):

- informed vigilance is actively promoted and rewarded;
- there exists an understanding of the entire system, its challenges and limitations is promoted and actively maintained;
- effective, real-time treatment process control, based on understanding critical capabilities and limitations of the technology, is the basic operating approach;
- fail-safe multi-barriers are actively identified and maintained at a level appropriate to the challenges facing the system;
- close calls are documented and used to train staff about how the system responded under stress and to identify what measures are needed to make such close calls less likely in future;
- operators, supervisors, lab personnel and management all understand that they are entrusted with protecting the public's health and are committed to honouring that responsibility above all else;
- operational personnel are afforded the status, training and remuneration commensurate with their responsibilities as guardians of the public's health;
- response capability and communication are improved, particularly as post 9–11 bioterrorism concerns are being addressed; and
- an overall continuous improvement, total quality management (TQM) mentality pervades the organisation.

9.4 CLOSING REMARKS

It is our hope that this text will go some way to preparing postgraduate water engineers, scientists and other interested professionals for their personal and organisational risk management journeys. If you have comments, suggestions for improvements or aspects of the book you would like to discuss, please do contact us at: s.pollard@cranfield.ac.uk.

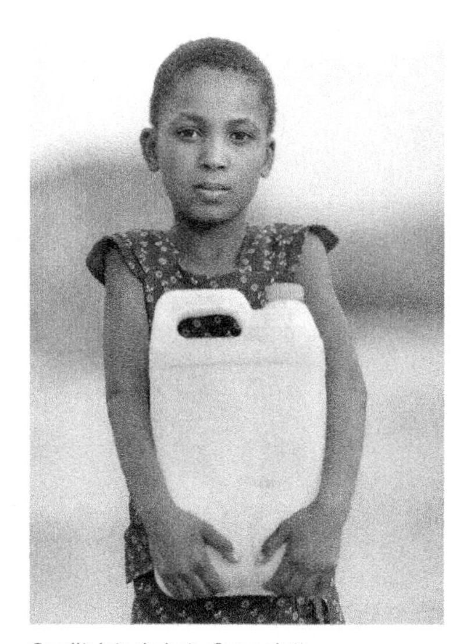

Credit: istockphoto © ranplett

Unit 10

Self assessment, abbreviated answers

This unit includes two sets of self-assessment questions and answers. In the section 10.1 below you will find abbreviated model answers to the questions in the main body of the text, chapter by chapter. Immediately below are a set of multiple choice questions to text your basic level competency in the issues raises in this book. The answers appear at the very end of this Chapter.

Risk and reliability engineering – questions

A total of 60 marks are available for 20 multiple choice questions. There is only one correct answer for each question. For some questions, one answer will be preferred over others.

Q	Marks	**Why manage risk?**				
1	2	*Knowing your organisation's risk appetite is critical if staff are to take measured risks within their job. Risk appetite is closely related to which other term?*				
	Ans:	*a hazard identification*	*b control loop*	*c risk acceptability*	*d ALARP*	*e human error*
2	2	*'Problem definition' is central to good risk management and precedes any risk analysis. It can be summarised by which of the following phrases?*				
	Ans:	*a 'know your risk'*	*b 'risk of what to whom?'*	*c rank your risks*	*d enterprise risk management*	*e frequency x consequences*
3	3	*Risk assessments are employed, usually for marginal situations, where the significance of the risk is uncertain. What is the primary purpose of the risk assessment?*				
	Ans:	*a to satisfy the regulator*	*b characterise the risk and inform risk management*	*c to stay out of the courts of law?*	*d to quantify uncertainties using probabilistic techniques*	*e to educate staff about health and safety*
4	3	*A service 'outage' is often a major issue for utilities. Considered as the end result of a series of preceding events, and not considering the harm that might result, is an outage a:*				
	Ans:	*a hazard?*	*b consequence?*	*c risk?*	*d probability?*	*e impact*
5	3	*In reviewing options for risk management, alongside considering the significance of the risk and the costs of implementing risk management measures, it is also essential to consider*				
	Ans:	*a the risk of what to whom*	*b the technology, management arrangements and social issues*	*c the regulatory system and expertise of officials*	*d the historic investment strategy within your organisation*	*e the nature of the hazard and human exposure to it*
Q	**Marks**	Basic probability and statistics				
6	4	*In a batch of 10 bearings, 3 are defective. Two bearings are selected randomly from the batch and installed in a pump. What is the probability that the pump will work properly?*				
	Ans:	*a 0.95*	*b 4.7*	*c 0.047*	*d 0.30*	*e 0.47*
7	5	*A component with an exponential life distribution has an MTTF (mean time to failure) of 30 000 hours. Calculate the probabilities of the following events (questions 7, 8, 9):* *(i) the component will survive 1 year of continuous service without failure*				
	Ans:	*a 0.08*	*b 3.708*	*c* 10^{-7}	*d 1/30 000*	*e 0.75*
8	5	*(ii) the component will fail between the end of the 4th and the end of the 5th year.*				
	Ans:	*a 0.08*	*b. 4.5*	*c 0.2077*	*d 0.75*	*e 0.25*
9	6	*(iii) the component will fail within a year, given that it has survived until the end of the 4th year*				
	Ans:	*a 0.25*	*b 4.5*	*c 0.08*	*d 0.95*	*e 0.0311*
Q	**Marks**	Process risk and reliability analysis				
10	2	*'ARM' is a term referring to the availability, reliability and maintainability of assets. The probability that an asset is in a state to perform its required function is its:*				
	Ans:	*a reliability?*	*b maintainability?*	*c availability?*	*d hazard state*	*e asset management*
11	3	*You are concerned about the failure of a major pump at your plant. The means by which it might break down are unknown to you. Which of the following risk tools and techniques will be of most value to you?*				
	Ans:	*a risk ranking*	*b HACCP*	*c event tree analysis*	*d HAZOP*	*e fault tree analysis*

Q	Marks		a	b	c	d	e
12	3		*Your in-plant wastewater treatment plant has failed and you are required to evaluate the impact of unlawful discharges to a sensitive surface water body downstream. Which of the following risk tools and techniques will be of most value to you?*				
		Ans:	*a HACCP*	*b event tree analysis*	*c fault tree analysis*	*d critical path analysis*	*e risk ranking*
13	2		*The multi-barrier concept:*				
		Ans:	*a can be used to describe active and latent flaws in unit processes in series*	*b is most useful for the analysis of parallel networks*	*c is a security measure for the prevention of unwanted site visitors*	*d describes microbial growth in water treatment plants*	*e is a novel disinfection process for oocyst removal in raw water*
14	3		*Consider a 4 component system, with each component having a reliability of 0.75. The overall reliability of the system is*				
		Ans:	*a 3*	*b 4.75*	*c* $3\sqrt{0.75}$	*d 0.3*	*e* $(4)^{0.75}$
Q	**Marks**		**Risks beyond the process boundary**				
15	3		*The source-pathway-receptor model is widely used for qualitative risk analysis and to inform the use of more quantitative risk analysis techniques. Which one of the following concepts is most central to this approach?*				
		Ans:	*a 'the dose makes the poison'*	*b first estimate the risk of what to whom*	*c without connectivity of the source, pathway and receptor, there can be no risk*	*d the dose-response relationship*	*e the characterisation of the risk*
13	3		*Which of the following characteristics of a hazard has most influence on the magnitude of a risk?*				
		Ans:	*a nature*	*b potency*	*c harm posed*	*d concentration*	*e timing*
17	2		*Under stress, human reliability can:*				
		Ans:	*a increase or fall, depending on the level of stress*	*b fall quickly resulting in cascading error*	*c usually be maintained indefinitely at stable levels*	*d be a reliable predictor of technological failure*	*e be estimate using quantitative probability techniques to improve systems reliability.*
18	2		*Reason's 'Swiss cheese' model is used to summarise:*				
		Ans:	*a the modes of failure for engineering processes*	*b environmental risk factors that, if not controlled, result in the loss of regulatory confidence*	*c organisational structures from strategic to operational risk*	*d physical deteriorations in the engineering structures of utility assets*	*e active and latent flaws within organisations and their contributions to incidents*
Q	**Marks**		**Risk and regulation**				
19	2		*In regulating utility companies, regulators generally seek:*				
		Ans:	*a confidence that process facilities can be operated safely and responsibly*	*b to prosecute the owners of systems that fail*	*c quantitative risk assessments from consultants*	*d to ensure compliance with ISO 14001 and related management systems*	*e to permit operations to continue without undue interference.*
Q	**Marks**		**Business risk management and implementation**				
20	2		*High reliability organisations have an active culture of:*				
		Ans:	*a preventative risk management*	*b risk management frameworks*	*c change management*	*d opportunity management*	*e organisational learning*

Total of 60 marks available for 20 questions

10.1 WHY MANAGE RISK?

Most set questions are penned very carefully. You should use the question to show what you know about the question, perhaps why it is asked and the context. Use every opportunity to do this, and be sure to answer the specific question being posed. Pay particular attention to the terminology used in the question. If it says 'list', then make a list; if it says, 'illustrate with an example', then use one.

SAQ 1.1 Think about the game of Monopoly™. Describe the consequences of landing on 'Mayfair', with and without a hotel on it, in risk terms. Use the terms probability and consequence in your answer.

The question is designed to test your understanding of the likelihood and consequences of a familiar adverse event – here of landing on the property Mayfair, that if owned by another player of the game, incurs a substantive penalty. Players within two dices throws (up to 12 moves) have a greater probability of those at distance. In Monopoly™, the consequences of landing on 'Mayfair' are dramatically increased if the owner has hotels on the property. There is also a bypass route. One of the 'Chance' cards directs the lucky recipient to go to 'Mayfair' directly! Being able to recognise the probability and consequence components of risk problems is an important first step towards risk management. Both components can be managed.

SAQ 1.2 A current concern is one of a bioterrorist attack on a water treatment works. What are the sources of a potential attack? What are the receptors? Describe the pathway – the mechanism by which such exposure may occur. Present the pathway as a chain of numbered events.

A further skills is to be able to identify the source, pathway and receptor components of a risk problem. Consequences can not be realised unless the receptor is exposed to the hazardous substance, or unless the chain of interim events occurs, following an initiating event, such that a hazardous situation is realised. In the above example, a critical aspect is access to the works. The sources of the hazard are the individuals that might plan such an attack and the hazardous agents that might be employed. The chain events can therefore be presented as an ordered list of interim events.

SAQ 1.3 Potent toxic chemicals reaching an activated sludge plant through a trade effluent discharge may render the microorganisms on which treatment relies ineffective. What is at risk in this example? How would you manage this risk – would you address the source, pathway or receptor? Draw a schematic of this situation before and after risk management.

The receptor here is the works and specifically, the biological treatment capacity of the secondary treatment plant. So the 'risk of what to whom' is the risk of the activated sludge plant being incapacitated by a shock load of trade effluent containing concentrations of hazardous compounds toxic to the activated sludge. Examples of such a situation might include the release of cyanide wastes through the foul sewer to the works. These risks are managed by trade effluent control which is often managed by the water utility. Consent limits are placed on trade effluent destined for the public treatment works. The receptor can also be managed – many activated sludge plants operating in urban conurbations become acclimatised to certain chemicals in effluent. Usually, however, attempts are made to manage these risks at source and often, in-plant treatment works are required so that large dischargers are responsible for their own effluent within their own company boundaries.

SAQ 1.4 What hazards might affect (a) an urban raw water intake; (b) an upland catchment reservoir?

The question is designed to test your ability to distinguish between hazard and risk. You might also use it to demonstrate your knowledge of point and diffuse releases. A good answer would first define a hazard: a substance or situation with the potential to cause harm. You might then list some examples of urban and rural hazards. Urban: combined storm overflows; releases of oils or chemicals following a transportation accident. Upland catchments: cryptosporidium and other pathogens from agricultural land; sulphur dioxide from acid rain; nitrogen and phosphorus from agricultural run-off. You might finish with a brief comment of the management of these hazards.

SAQ 1.5 On a scale of 1 to 5, qualitatively rank the following adverse consequences of water treatment failures and justify your answer:

- *an E. Coli 0157:H7 (potent human pathogen) outbreak;*
- *the presence of benzene in drinking water at concentrations below the drinking water standard;*
- *musty tasting water from the tap.*

Note down the uncertainties that make your ranking difficult to complete.

Here you are being tested on your knowledge of the potential harm that could be realised if the source, pathway and receptor became connected and exposure occurred. Again, it is good to show you understand what the question is getting at and briefly state what you understand by harm. A further dimension here is the relative potency of these various hazardous agents, should exposure occur. The questioner wants to know that you understand that water treatment failures usually provide little time to intervene. The question assumes exposure has occurred, so you are left with ranking the consequences. Remember for these exercises you must define your scale (1–5) and justify your ranking of each consequence, here by reference to the potency of the hazardous agent in each case. E Coli 0157:H7 is a potent pathogen; benzene, though an inhalation carcinogen, at concentrations below a drinking water standard is significant but unlikely to be acutely serious; a musty tasting water supply, though aesthetically unpleasant, is likely to be temporary. You might rank the consequences of these exposures as 5, 3, 1 respectively where 5 is serious. But qualify your answer – what assumptions have you made? No risk analysis is complete without this.

SAQ 1.6 "I've heard plasticizers leach into bottled water over time and may be affecting my health. Should I be concerned?" Pen a brief answer to this customer who has called your customer helpline.

This is challenging. It requires you to be able to comment on exposure and the impact of phthalates. The customer seeks reassurance and confidence and you need to provide a response that presents our current state of knowledge. Remember that exposure (dose) is concerned with both amount (concentration) and duration (t). Whilst there are concerns about phthalate exposures, especially for new born infants, the risks from these compounds affecting the general population is believed to be close to zero.

10.2 BASIC STATISTICS AND PROBABILITY

Quantitative questions require you to illustrate that you understand the premise, the appropriate equation and that you can then progress with the intermediate steps to gain a correct numerical answer.

SAQ 2.1 If the probability of failure of an alarm device is p, calculate the probability that at least one out three alarms will fail.

This question is about redundancy in alarm systems. If you have time, you might state that you understand the context of the question and illustrate you answer with an example. Here, since the probability that none of the devices will fail is (1–p)3, the probability that at least one of the devices will fail is 1–(1–p)3.

SAQ 2.2 Assume that the probability of a pathogen infecting an exposed individual is p. If n pathogens are ingested and they infect independently of one another, calculate the probability of infection.

Here, infection occurs when at least one pathogen 'succeeds'; that is, it infects. The probability that none of the pathogens will infect is $(1-p)^n$. *The probability that at least one of the pathogens will infect is then* $1-(1-p)^n$, *which is identical to the solution above. In reality, when p is small, the probability of infection (P) can be approximated to* $1-e^{[-n \ln(1-p)]}$ *which is approximately equal to* $1-e^{(-np)}$, *since for small probabilities,* $\ln(1-p) \cong -p$.

SAQ 2.3 Why is preventative maintenance not appropriate for components displaying an exponential life distribution?

This test your understanding of the characteristics of the exponential distribution. Remember that the exponential distribution is 'memory-less'.

This means in practice that the probability (as described by an exponential distribution) that a component will work for a particular time is independent of how long it has worked; that is, it is as good as new. This clearly is a limited condition for maintenance, where component lives are, in part, determined by their length of service and condition.'

10.3 PROCESS RISK AND RELIABILITY ANALYSIS

SAQ 3.1 Consider an activated sludge treatment process. How do the terms 'reliability', 'availability' and 'maintainability' apply to this unit process?

This question is about the concepts fundamental to assessing the reliability of assets. There are two aspects to the question – do you understand these three terms and can you then apply them in a practical context? The terms are defined in section 3.1 of unit 3. The probability is the likelihood that the plant is an appropriate condition to perform secondary treatment; the availability usually relates to its downtime; the maintainability relates the ability to bring the plant back up to the desired availability following maintenance.

SAQ 3.2 Consider the equation in Example 3.1. How might this equation be amended to account for the time associated with the preventative maintenance (PM) of unit processes?

We now consider the quantification of availability and maintenance. Reliability studies are primarily concerned with the failure of components and systems. However, in many real world systems, failed components can be repaired or replaced thus restoring the system back in to the working state. This is particularly relevant to process systems or production systems. As in reliability analysis, the engineer is interested in determining the probability that the system is in the working or failed state but in this case it is necessary to consider the restoration rate as well as the failure rate over the full life cycle of the system.

Availability is defined as the probability that the system is in the working /operating state and is dependent on the failure rate and the repair rate. In the steady state, availability is independent of time and may be expressed mathematically (section 3.1) as:

$$A = \frac{\text{MTBF}}{\text{MTBF} + \text{MTTR}} = \frac{\text{MEAN\_UP\_TIME}}{\text{TOTAL\_INSTALLED\_TIME}}$$

where MTTF is the mean time between failure and MTTR is the mean time to restore. Availability then is the proportion of time that the system is in the operating state

since $\text{MTTF} = 1/\lambda$

and $\text{MTTR} = 1/\mu$

Preventative maintenance seeks to retain the system in an operational state for longer by preventing failures from occurring by servicing the asset (cleaning, lubrication etc.). This affects the availability of the asset because it must be taken off line. It introduces a further denominator to the equation above.

$$A = \frac{\text{MTBF}}{\text{MTBF} + \text{MTTR} + \text{MPM}} = \frac{\text{MEAN\_UP\_TIME}}{\text{TOTAL\_INSTALLED\_TIME}}$$

Where MPM is the mean preventative maintenance time. It can be seen that MPM must be kept to a minimum to ensure the availability is kept high.

10.4 ASSESSING RISKS BEYOND THE UNIT PROCESS BOUNDARY

SAQ 4.1 From source, through treatment, on to distribution and consumer supply, describe individual barriers that contribute to the management of pathogenic risk in drinking water.

The multi-barrier approach to risk management for drinking water is well-established. Here are some of the protective barriers that act to minimise a pathogenic challenge. Note they do not just refer to unit processes for drinking water treatment. They include (i) preventing contamination of the sources of water, through source control (including land management); (ii) using appropriate water treatment (filtration, coagulation) processes and distribution systems; (iii) water testing (maintenance of the chlorine residual); and (iv) training of water managers. Following the Walkerton incident,, Justice O'Connor called for a multiple-barrier water management approach to prevent a similar tragedy from occurring again. He concluded that source protection is the most effective and efficient means of protecting the safety of drinking water.

SAQ 4.2 Undertake a DPSIR analysis for the impact of intensive agriculture on nitrate loadings to groundwaters.

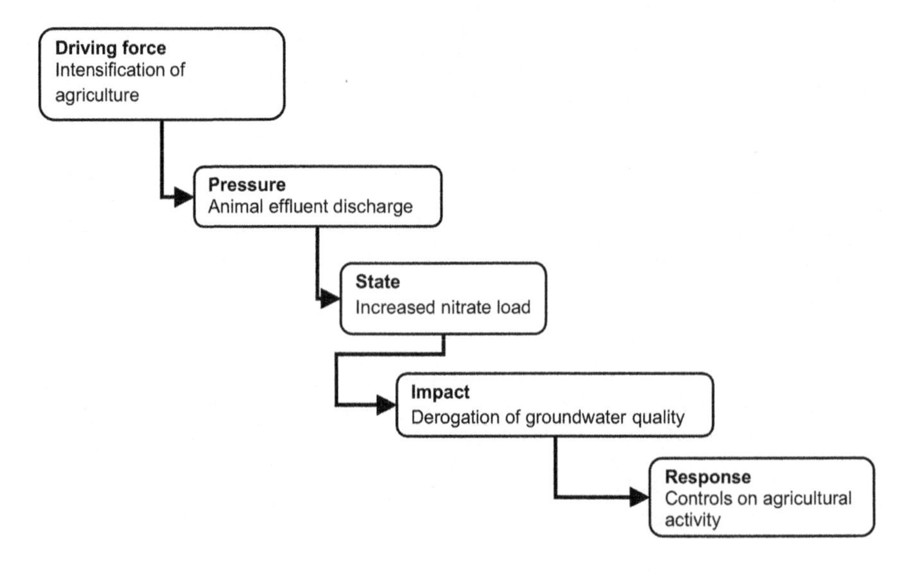

SAQ 4.3 What shortcomings does the mathematical modelling of human error embody?

Human reliability analysis (HRA) treats individuals as components in a chain with an associated reliability. However, the models produced can not predict human behaviour though they may attempt to characterise typical failure rates. Good HRA models incorporate the factors that allow for the operating conditions that may result in an error. These models still have shortcomings however in that they: (a) tend to focus on individual, rather than collective behaviours and the social interactions that inform them; (b) have yet to be fully validated; (iii) tend to focus on equipment operators rather than individuals, organisational policies or company rules higher in the organisational hierarchy. See: Redmill, F. (2002) The significance to risk analysis of risks posed by humans, Engineering Management Journal, 12(4) August 2002.

SAQ 4.4 Compare and contrast the principles of management control for hazards in (a) catchments; (b) distribution systems.

This requires you to compare and contrast risk analysis and management for these two types of problem. Catchments are characterised by being open systems, having a complex numbers of point and diffuse sources of a hazard and needing voluntary codes of good practice alongside consented discharges to ensure that activities such as land management, agricultural practice, pesticide applications and forestry management take account of risks to water sources. Distribution systems are (usually) closed engineered systems, amenable to Markov modelling and network analysis. Here controls relate to reducing the probability of pipe bursts and sewer collapses.

10.5 REGULATING WATER UTILITY RISKS

SAQ 5.1 Consider yourself to be a regulator of a large sewage treatment works. What specific evidence would you seek as assurance of good risk management? Other than consent records, what would you look for during an inspection of the works?

Regulatory inspections are likely to include inspections of:
- *screenings and grit removal*
- *the physical condition of storm water tanks*
- *odour control and management*
- *procedures for sludge disposal*
- *adequacy of chemical storage tanks and their bunding*

SAQ 5.2 What risk management advice would you offer farmers to prevent spray drift to surface water bodies. Include advice that addresses both the probability and consequences of undesirable ecological impacts.

The question tests your knowledge of the role of codes of good practice. The PEPFAA code includes the following advice on spray drift:

Don't spray crops unless the weather conditions are right (wind speed and direction may increase the probability of drift landing on water courses)

Don't spray crops without selecting the nozzle system to suit the product(s) being applied, the crop and spray volume (inappropriate nozzle may increase the probability of over application or misdirected spray)

Don't permit spray or spray drift to endanger sensitive habitats (consequences of impacting sensitive ecological receptors)

Don't neglect routine maintenance and calibration of spray equipment (increases the likelihood of over application)

10.6 CORPORATE RISK GOVERNANCE

SAQ 6.1 Propose a structure for the governance of risk within a water utility. Suggest the regularity of reappraising risk assessments for each part of the business.

The question tests your understanding of corporate arrangements for risk management. You may refer to the risk hierarchy and governance structures in Figures 6.2 and 6.3. Your answer should include the following essential components: (i) an internal audit (financial control) function; (ii) Board level accountability for risk management (Chief Financial Officer, Finance Director); (iii) risk management committee reporting to the executive; (iv) group risk manager and risk champions. Strategic, tactical and operational risks might be formally appraised quarterly, 6-monthly and 6annually, respectively.

SAQ 6.2 Think about your own appetite for risk. How does this impact on you seeking new opportunities? Now consider what parallels there are for utilities seeking to (i) adopt a new technology e.g. membrane treatment; (ii) acquire a new water company in Asia. What due diligence might you adopt before making these decisions?

One's risk appetite depends on one's values and one's perceptions of the rewards of risk-taking. In corporate terms, this will often be translated into the amount of manageable business interruption losses that an organisation could withstand. However, risks that are less quantifiable are more challenging to appraise. New technology – risks of perturbations to an existing stable system and increased uncertainty of performance vs. reduce treatment costs and increased regulatory compliance. New company acquisition – risks of uncertain market conditions and performance vs benefits of greater market penetration. Reducing the uncertainty in these decisions can be achieved by practical R&D and pilot-scale treatability studies and a full liability assessment.

SAQ 6.3 List 5 risks the water utility sector will need to actively manage considering the demographic changes in the workforce population. How might the sector manage these risks?

(1) Loss of in-company expertise amongst older staff
(2) Loss of historical corporate knowledge, especially with respect to incidents and their management
(3) Different personal values amongst new employees – same levels of vigilance?
(4) More part-time workers – potential loss of continuity
(5) Increased training requirements

10.7 MANAGING OPPORTUNITIES AND REPUTATIONS

SAQ 7.1 List 4 catchment level risks that can be managed. Now consider what opportunities might be secured from managing these risks.

(1) Reduced construction site run-off (better site management and less mud on approach roads)
(2) Reduced groundwater pollution (long term sustainability of groundwater resources)
(3) Reduced run-off through implementation of sustainable urban drainage systems (increased amenity value of flow equalisation ponds, grassed areas)
(4) Reduced soil erosion (retention of organic carbon)

SAQ 7.2 Undertake a PEST analysis for the adoption of tertiary (granular activated carbon) treatment of wastewater from a sewage treatment works.

Political		***Economic***	
Risk	*Opportunity*	*Risk*	*Opportunity*
• Changing regulations • Potential regulatory climbdown	*• Early adopter – regulatory approval*	*• Costs of implementation • Costs of carbon regeneration*	*• Reduced compliance costs • Possible increased investment*
Social		***Technological***	
Risk	*Opportunity*	*Risk*	*Opportunity*
• Public perceptions of GAC tanks	*• Reputational enhancement • Potential CSR advantages*	*• First to market – first to discover process challenges*	

SAQ 7.3 Devise a scheme for categorising water quality incidents. Suggest 3 attributes to define the incident and then 3 levels of incident – moderate, major and serious.

Regulators are increasingly concerned with categorising water quality incidents by reference to risk so that they can resource incident investigations accordingly and proportionately to the risks posed. The Environment Agency's 1995–1998 incident classification scheme had 4 categories, from 1 (major) to 4 (unsubstantiated) and included the following attributes:

Definition of pollution incident categories – used from 1995 – 1998

Category 1

A major incident involving one or more of the following:

- (a) *potential or actual persistent effect on water quality or aquatic life;*
- (b) *closure of potable water, industrial or agricultural abstraction necessary;*
- (c) *extensive fish kill;*
- (d) *excessive breaches of consent conditions;*
- (e) *instigation of extensive remedial measures;*
- (f) *significant adverse effect on amenity value;*
- (g) *significant adverse effect on site of conservation importance.*

Category 2

A significant incident involving one or more of the following:

- (a) *notification of abstractors necessary;*
- (b) *significant fish kill;*
- (c) *readily observable effect on invertebrate life;*
- (d) *water unfit for stock watering;*
- (e) *bed of watercourse contaminated;*
- (f) *amenity value to downstream users reduced by odour or appearance.*

Category 3

A minor incident resulting in localised environmental impact only. Some of the following may apply:

- (a) *notification of abstractors not necessary;*
- (b) *fish kill of less than 10 fish (species of no particular importance to the affected water);*
- (c) *no readily observable effect on invertebrate life;*
- (d) *water not unfit for stock watering;*
- (e) *bed of watercourse only locally contaminated;*
- (f) *minimal environmental impact and amenity value only marginally affected.*

Category 4 (unsubstantiated)

A reported pollution incident that upon investigation proves to be unsubstantiated, that is, no evidence can be found of a pollution incident having occurred.

10.8 EMBEDDING BETTER DECISION-MAKING WITHIN UTILITIES

SAQ 8.1 What do you understand by the term 'preventative risk management'? Provide an example for each of the strategic, programme (tactical) and operational contexts of water utility management.

Preventative risk management means putting in place measures, controls and behaviours that act to prevent incidents before they occur.

This is important in the water and wastewater utility sector because the opportunities to intervene and minimise the consequences of failure are very limited. One example of preventative risk management is the adoption of drinking water safety plans that aim to identify and manage the critical control points in the drinking water supply chain from catchment to tap.

SAQ 8.2 Why is knowledge management important to effective risk management?

Risks are dynamic. Because they are determined in space and time, they are continually changing. Thus, company risk profiles and the risk status of individual treatment works is in continual flux. Organisations that can manage their risk information and convert 'snapshots' of risk into meaningful dynamic risk profiles are able to verify and validate the value of their risk management activity.

SAQ 8.3 What kinds of company 'stories' might be useful in helping generate a risk management culture? Can you devise one? Ensure is includes 3 lessons learnt.

Corporate stories about adverse situations that have happened, risks that were taken and paid off, close calls or lessons learnt are all important 'cultural items'. They personalise the consequences of incidents or of opportunistic successes and make them more real to people. The types of stories that are valuable in organisations might include those regarding (i) health and safety incidents; (ii) successful secondments to risk teams; (iii) rewards for innovative risk management suggestions; (iv) chronological accounts of calculated risks and the benefits that flowed from these decisions.

Q	Marks	**Why manage risk?**				
1	2	*Knowing your organisation's risk appetite is critical if staff are to take measured risks within their job. Risk appetite is closely related to which other term?*				
	Ans:	*a hazard identification*	*b control loop*	***c risk acceptability***	*d ALARP*	*e human error*
2	2	*'Problem definition' is central to good risk management and precedes any risk analysis. It can be summarised by which of the following phrases?*				
	Ans:	*a 'know your risk'*	***b 'risk of what to whom?'***	*c rank your risks*	*d enterprise risk management*	*e frequency x consequences*
3	3	*Risk assessments are employed, usually for marginal situations, where the significance of the risk is uncertain. What is the primary purpose of the risk assessment?*				
	Ans:	*a to satisfy the regulator*	***b characterise the risk and inform risk management***	*c to stay out of the courts of law?*	*d to quantify uncertainties using probabilistic techniques*	*e to educate staff about health and safety*
4	3	*A service 'outage' is often a major issue for utilities. Considered as the end result of a series of preceding events, and not considering the harm that might result, is an outage a:*				
	Ans:	*a hazard?*	***b consequence?***	*c risk?*	*d probability?*	*e impact*
5	3	*In reviewing options for risk management, alongside considering the significance of the risk and the costs of implementing risk management measures, it is also essential to consider*				
	Ans:	*a the risk of what to whom*	***b the technology, management arrangements and social issues***	*c the regulatory system and expertise of officials*	*d the historic investment strategy within your organisation*	*e the nature of the hazard and human exposure to it*
Q	**Marks**	**Basic probability and statistics**				
6	4	*In a batch of 10 bearings, 3 are defective. Two bearings are selected randomly from the batch and installed in a pump. What is the probability that the pump will work properly?*				
	Ans:	*a 0.95*	*b 4.7*	*c 0.047*	*d 0.30*	***e 0.47***
7	5	*A component with an exponential life distribution has an MTTF (mean time to failure) of 30 000 hours. Calculate the probabilities of the following events (questions 7, 8, 9):* *(i) the component will survive 1 year of continuous service without failure*				
	Ans:	*a 0.08*	*b 3.708*	*c 10^{-7}*	*d 1/30 000*	***e 0.75***
8	5	*(ii) the component will fail between the end of the 4th and the end of the 5th year.*				
	Ans:	***a 0.08***	*b. 4.5*	*c 0.2077*	*d 0.75*	*e 0.25*
9	6	*(iii) the component will fail within a year, given that it has survived until the end of the 4th year*				
	Ans:	***a 0.25***	*b 4.5*	*c 0.08*	*d 0.95*	*e 0.0311*
Q	**Marks**	**Process risk and reliability analysis**				
10	2	*'ARM' is a term referring to the availability, reliability and maintainability of assets. The probability that an asset is in a state to perform its required function is its:*				
	Ans:	*a reliability?*	*b maintainability?*	*c **availability**?*	*d hazard state*	*e asset management*
11	3	*You are concerned about the failure of a major pump at your plant. The means by which it might break down are unknown to you. Which of the following risk tools and techniques will be of most value to you?*				
	Ans:	*a risk ranking*	*b HACCP*	*c event tree analysis*	*d HAZOP*	***e fault tree analysis***

Q	Marks					
12	3	*Your in-plant wastewater treatment plant has failed and you are required to evaluate the impact of unlawful discharges to a sensitive surface water body downstream. Which of the following risk tools and techniques will be of most value to you?*				
	Ans:	*a HACCP*	*b* ***event tree analysis***	*c fault tree analysis*	*d critical path analysis*	*e risk ranking*
13	2	*The multi-barrier concept:*				
	Ans:	*a* ***can be used to describe active and latent flaws in unit processes in series***	*b is most useful for the analysis of parallel networks*	*c is a security measure for the prevention of unwanted site visitors*	*d describes microbial growth in water treatment plants*	*e is a novel disinfection process for oocyst removal in raw water*
14	3	*Consider a 4 component system, with each component having a reliability of 0.75. The overall reliability of the system is*				
	Ans:	*a 3*	*b 4.75*	*c* $3\sqrt{0.75}$	*d* ***0.3***	*e* $(4)^{0.75}$
Q	**Marks**	**Risks beyond the process boundary**				
15	3	*The source-pathway-receptor model is widely used for qualitative risk analysis and to inform the use of more quantitative risk analysis techniques. Which one of the following concepts is most central to this approach?*				
	Ans:	*a 'the dose makes the poison'*	*b first estimate the risk of what to whom*	*c* ***without connectivity of the source, pathway and receptor, there can be no risk***	*d the dose-response relationship*	*e the characterisation of the risk*
13	3	*Which of the following characteristics of a hazard has most influence on the magnitude of a risk?*				
	Ans:	*a nature*	*b* ***potency***	*c harm posed*	*d concentration*	*e timing*
17	2	*Under stress, human reliability can:*				
	Ans:	*a* ***increase or fall, depending on the level of stress***	*b fall quickly resulting in cascading error*	*c usually be maintained indefinitely at stable levels*	*d be a reliable predictor of technological failure*	*e be estimate using quantitative probability techniques to improve systems reliability.*
18	2	*Reason's 'Swiss cheese' model is used to summarise:*				
	Ans:	*a the modes of failure for engineering processes*	*b environmental risk factors that, if not controlled, result in the loss of regulatory confidence*	*c organisational structures from strategic to operational risk*	*d physical deteriorations in the engineering structures of utility assets*	*e* ***active and latent flaws within organisations and their contributions to incidents***
Q	**Marks**	**Risk and regulation**				
19	2	*In regulating utility companies, regulators generally seek:*				
	Ans:	*a* ***confidence that process facilities can be operated safely and responsibly***	*b to prosecute the owners of systems that fail*	*c quantitative risk assessments from consultants*	*d to ensure compliance with ISO 14001 and related management systems*	*e to permit operations to continue without undue interference.*
Q	**Marks**	**Business risk management and implementation**				
20	2	*High reliability organisations have an active culture of:*				
	Ans:	*a* ***preventative risk management***	*b risk management frameworks*	*c change management*	*d opportunity management*	*e organisational learning*
Total of 60 marks available for 20 questions						

Lightning Source UK Ltd.
Milton Keynes UK
UKOW07f2042040816

279989UK00004B/14/P